S0-BZI-925

Bioethics as Practice

STUDIES IN

SOCIAL MEDICINE

Allan M. Brandt &

Larry R. Churchill,

editors

Bioethics as Practice

Judith Andre

The University of North Carolina Press

Chapel Hill and London

Manufactured in the United States of America
Set in Cycles type by Tseng Information Systems
The paper in this book meets the guidelines for permanence and durability of the Committee on Production Guidelines for Book Longevity of the Council on Library Resources.

Library of Congress
Cataloging-in-Publication Data
Andre, Judith.
Bioethics as practice / Judith Andre.
p. cm. — (Studies in social medicine)
Includes bibliographical references and index.
ISBN 0-8078-2733-9 (cloth : alk. paper)
1. Medical ethics. 2. Bioethics.
[DNLM: 1. Bioethics. 2. Bioethical Issues.
QH 332 A555b 2002] I. Title. II. Series.
R724 .A663 2002
174'.2—dc21 2002001544

06 05 04 03 02 5 4 3 2 1

FOR PAT

Contents

Preface

In 1991, after eighteen years of teaching philosophy in traditional academic settings, I took a position in bioethics at Michigan State University. I found the transition more difficult than I had expected. I had moved before, far too often; I belong to the academic generation known as "gypsy scholars," forced by the lack of jobs to move from state to state in pursuit of work. During the 1970s I taught at seven different colleges and universities in six cities (and four states) before finding a tenure-stream position at Old Dominion University. I had found those seven institutions fundamentally alike. Moving so often, and such distances, was hard, but I knew what I had to do to begin again: find a place to live, reach out for friends, and find out where the chalk was kept and how to use the photocopier.

The move to Michigan State should have been easier than the earlier ones. For the first time, I was moving voluntarily, leaving a tenured position and choosing between two offers elsewhere; and in a sense I was coming home, as I had done my graduate work here. I was joining a unit with national stature and a set of enviably fine colleagues. Part of the unexpected pain came from the fact that my ten years at Old Dominion had been so rewarding—for the first time, moving meant leaving behind something solid and good. But part of the problem was also culture shock. There were deeper things to learn than where the chalk was kept. Although I had taught bioethics to undergraduates and done several fellowships in the field, there was virtually nothing in my new occupation that corresponded to what I had been doing for so long. The difference was not primarily a matter of knowledge (I did have a lot to learn, but I knew how to study). The difference was practical: the structure of the courses in which I taught, the forums for which I wrote, and the ways in which I served the community were all new.

It took me several years to realize how deep the differences were, to understand how much I had to learn that could not be found in books. I began to see that this new form of professional life demanded skills that have nothing to do with academic philosophy. (Ethics consultations within a hospital, for instance, often require mediation skills.) Soon I began to see this new form of professional life as an example of what Alasdair MacIntyre calls a practice, a complex activity evolving over time, whose aspects can only be understood in terms of the whole. (His examples include architecture, portrait painting, and chess.) Such practices cannot survive without virtues, and this analysis fits with my sense that my new life presented new moral demands.

I set out to understand my new situation. The quest involved reading, reflection, and talking to other people. Beginning in 1996, I visited more than fifteen ethics centers, most in bioethics, but some in related fields like business or professional ethics. I talked with about seventy-five people, some at the centers I visited, others at professional conferences. I looked for people who were situated differently from me and from each other. I promised everyone confidentiality, and so the stories in this book are altered and disguised; the more negative the point, the more effort I put into disguising those involved. I did occasionally want to give credit for someone's insight, turn of phrase, or accomplishment, and so hit upon the old strategy of hiding like with like: Some of the people named within the book are people I interviewed, but most are not. And conversely, some of the people I interviewed occur by name in what I've written, but most do not. Although most of the quotations come from my formal interviews, a number come from more casual conversations.

I have struggled with a question of basic terminology, with the right name for the field I was describing. I knew that I did not want to use "medical ethics," since the word "medicine" properly refers to the work of physicians, and even if I were only talking about health care (and not health policy and the life sciences), I would avoid a term that rendered nurses, therapists, social workers, and chaplains invisible. But "bioethics" also seemed too narrow. For one thing it suggests a specific, technical sort of work, the activity of concluding that some choices are, and others are not, ethically right. What we do, in fact, is much broader; it includes every kind of examination of the moral dimensions of health care, health policy, the biological sciences, and cultural stances toward health and sickness. Our work necessarily involves most of the humanities: philosophy, law, literature, the social sciences, and religious studies. But every phrase that tries to capture that fact is so ungainly as to be useless. My own workplace, for instance, is called the Center for Ethics and Humanities in the Life Sciences. One of the original professional associations for the field was called the Society for Health and Human Values; the present national association is called the American Society for Bioethics and the Humanities. All of these are generally referred to with acronyms or shortened phrases. I could do no better, so in the end I settled on "bioethics" as the least unsatisfactory term that was still usable. I hope the book will make clear how broadly I mean the term.

I am immensely grateful to Larry Churchill, who invited me to write this book, and to everyone who talked with me about their lives in bioethics and the medical humanities. (At least here I can acknowledge the fact that some would reject the label "bioethics" for their work.) I am grateful as well to those who provided me with many kinds of working space: John Fletcher and Ed Spenser at the University of Virginia; Dorothy Vawter and Karen Gervais at the Minnesota Center for Health Care Ethics; and Mary Mahowald, Mark Siegler, and the other faculty and fellows of the McLean Center for Medical Ethics at the University of Chicago. My thinking on the subject of moral development began during a fellowship in the Harvard University Program in Ethics and the Professions. I am most indebted, however, to my colleagues at Michigan State who welcomed me in 1991 and supported me throughout the years this book took shape: Howard Brody, Len Fleck, and Tom Tomlinson, each so well known that most readers will recognize how fortunate I am to be working with them; Jan Holmes, our secretary, for ten years an indispensable help in everything I have done. Others who joined the center after me have been at least as loyal and generous: Clayton Thomason, lawyer and priest, whose collegiality has been a welcome addition; Libby Bogdan-Lovis, whose administrative assistance and anthropological perspective are invaluable; and Beth McPhail, an especially welcome addition to our office staff. In the last stages of writing, I was fortunate to have the assistance of Sonya Charles, a graduate student in philosophy at MSU, whose ability and diligence carried me through.

An earlier version of chapter 1 appeared as "The Week of November 7: Bioethics as a Practice," in *Philosophy of Medicine and Bioethics,* edited by Ron Carson and Chester Burns (Boston: Kluwer Academic Publishers, copyright 1997), 153–72; I am grateful for permission to include it in this work. Tod Chambers collaborated with me on earlier drafts of chapter 4. Part of chapter 6 is based on "The Medical Humanities as Contributing to Moral Growth and Development," in *Practicing the Medical Humanities: Forms of Engagement,* edited by Ronald Carson, Chester R. Burns, and Thomas R. Cole (Hagerstown, Md.: University Publishing Group, forthcoming). Some of chapter 8 first appeared as "Speaking Truth to Employers," *Journal of Clinical Ethics* 8:2 (Summer 1997): 125–29. The discussion of humility in chapter 10 is drawn from "Humility Reconsidered," in *Margin of Error: The Ethics of Mistakes in the Practice of Medicine,* edited by Susan B. Rubin and Laurie Zoloth (Hagerstown, Md.: University Publishing Group, 2000), 59–72.

1 November 1994

I am a faculty member in a unit called the Center for Ethics and Humanities in the Life Sciences (CEHLS). The awkwardness of the center's title reflects the uncertainty felt everywhere in the field—the field that for convenience I call bioethics[1]—about how to name it, how to describe what unifies work done by people from disparate backgrounds in a set of loosely related tasks. At the moment, for instance, the people I work with include philosophers, social scientists, physicians, nurses, a medievalist, a lawyer, and a priest. (A number of these people wear more than one hat.) Some have official appointments with CEHLS, but others are affiliated only in the sense that they and we work together on various projects. The projects, too, vary widely; they might concern our home institution (Michigan State University), local hospitals, the state legislature, or still wider arenas.

This book is, in part, an attempt to provide a unifying framework for all this. Its origins lie in a paper I wrote some years ago for Ron Carson and Chester Burns, who had invited me and a number of others in the field to a working conference in Galveston, Texas.[2] Because the conference marked a number of anniversaries, Carson and Burns asked participants to reflect on the progress of the field, to reflect upon their own careers, to think about the way the field had changed, lived up to its promise or failed to, met its goals or changed them. Although the conference officially focused on philosophy and medicine, inevitably the participants came from a wider set of backgrounds; furthermore, they had been deliberately chosen to represent various career paths and stages. I accepted with enthusiasm, partly for the chance to work out some ideas, partly for the chance to go back to Galveston. My life in bioethics had essentially begun there, during a six-month fellowship in 1990, when I first learned the pleasure of applying philosophy directly to life, and the freedom of working in an interdisciplinary context. I was eager to return.

True to its origins in the Galveston conference and to the character of the center in which I work, this book speaks in a variety of voices: personal reflection, allegory, moral argument, and philosophical analysis. It includes a variety of voices in a more literal sense as well; soon after writing the paper for the Galveston conference, I set out to talk with others in bioethics and the medical humanities, as many people and as differently situated as possible. I asked them what they did and what they thought about what they did, and their comments are an important part of this book.

In that now-distant November, however, I was still reflecting only on my own activities. The paper I wrote for Ron and Chester began with personal reflection, a description of one week in my professional life. Although the week of November 7, 1994, was typical in some ways, it also provided a unique focus for my paper: November 7 was election day, the off-year election following Bill Clinton's victory in 1992. In 1994 the Republicans swept into power, taking control of both houses of Congress. More personal, a difficult ethics consultation suddenly brought into focus how different my job now was from the traditional academic posts I had once held. So a number of themes, public and personal, converged.

The Week of November 7

A distinctive feature of our work in CEHLS is the amount of driving it demands. We serve two medical schools (since one is osteopathic, we refer to the other as allopathic), both part of MSU's enormous single campus, but there is no university hospital. Our medical students spend their first two years (what are called the classroom years) on campus, and then move for clinical training to community hospitals in the city and around the state. In addition, given the size of the East Lansing campus (thousands of acres), we often drive even to committee meetings or the library, more than a mile from our offices. So that week found me frequently in my car, darting around the campus and the city. And the week provided (as my time in Galveston had taught me to expect) many different ways to learn: reading, of course, but also listening, watching, and interacting. It also provided a variety of ways to act. On Monday the phone rang almost as soon as I arrived at my desk. Could I lead an ethics consult at a Lansing hospital at eleven o'clock? Someone else had done the initial work: received the physician's request, decided that it presented an ethical issue, decided the team should meet with the patient's daughter without the doctor, and arranged the time and place. My caller gave a brief overview of the situation, to which I listened with interest but detachment; I had learned before that the first, thumbnail sketch of these cases, usually a third-hand description, is not too useful. The real story is always more complex, and sometimes the real issue is different from what it seems at first. In this case I was told that a stroke patient's daughter was fighting the doctor over tube feeding, which she did not believe should be started.

I walked into the meeting room to find the daughter, whom I'll call Catherine Bactri, in tears, and being comforted by another person from the consultation team. A third member joined us, reporting that the patient had been told of the consult and demanded to be included. I felt we needed to talk with Catherine and with one another first, a decision about which I later had second thoughts. At any rate, we sat down with Catherine, a middle-aged woman, divorced, an only child, facing her mother's illness alone while holding down a full-time job. Her mother, whom I'll call Geneva Bactri, was eighty-eight and recovering from a stroke. The extent to which she would recover was unclear, and without a gastrostomy (a tube that allows food to be put directly into the stomach, through an opening in the abdominal wall) she would probably die soon. She often lapsed into unarousable sleep, but at times awoke and seemed to know what was going on. At those times she was adamant: "No gastrostomy. No stomach tube."

Everything about this situation was familiar: the issue (refusal of treatment); the process (we listened, we talked; we asked about how Mrs. Bactri had lived her life and what she had said she wanted); and the resolution (a patient or her surrogate has a right to refuse treatment, even when the refusal shortens her life). Catherine Bactri was frustrated and exhausted when we started: "I'm only trying to do what I always promised my mother I would." We listened with attention and respect, and when she burst out that the attending physician "made her feel terrible," reminded her quietly that she was free to change doctors. By the end of the meeting she was calm; she had been heard. Both the meeting and the report I wrote up were satisfying.

As I composed our report, I realized that as a graduate student in philosophy I would have been astonished to know that twenty years later I would be writing a report for a hospital chart. I also found myself wondering about what we had accomplished, and about the cost. Consultations are a slow and expensive process. Four professional people—doctor, nurse, philosopher, and biochemist—had each spent several hours on the case. If the doctor listened to us, Mrs. Bactri's life would probably be shortened slightly: a few months, perhaps, less than 1 percent of her eighty-eight years. That amount of time in itself, gained or lost, did not seem significant. On the other hand, Mrs. Bactri's daughter felt far better after the consultation, and perhaps that was the significant result. The physician probably felt more comfortable, too. I didn't know if he had asked for the consultation because he thought it would protect him legally, or because he was genuinely uncertain about the right thing to do; both motivations are common. In either case, I felt uneasy. On the one hand, the issues were not ethically complex, and the physician should have been able to sort things out for himself; on the other, he should not be using an ethics consultation as a form of legal protection. I worried, too, that we were making it easier for the doctor not to take the time really to talk with Catherine Bactri. I was somewhat consoled by the young family physician (an MSU graduate) who helped with the consultation; her unhurried, respectful attention to the patient was a model of what we try to teach.

Once back in my office, after this unscheduled three hours away, I turned to my own writing. Ten years before, I had been writing an article about the moral status of actions that affect no one except the doer, using technical tools from analytic philosophy. I submitted it to a number of philosophy journals, appreciated or resented the reviewers' comments, and eventually published it. I was unsure whether or not anyone had ever read it. This Monday, however, my topic was very small premature babies. I had been asked to write the piece for one of the newsletters my center publishes, and I knew first, that no referee would pass judgment on it, and second, that it would be printed and it would be read. It was less careful and less technical than what I was used to doing and not at all original—but it would probably make a difference in what people did.

The next day, election day, began for me with an 8:00 A.M. undergraduate bioethics class. I had taught undergraduate philosophy for almost twenty years before coming to MSU, but did so much less frequently now. Since teaching can be intensely rewarding but also deeply painful, I was grateful that other activities now buffered my engagements in the classroom.

On that Tuesday I needed all the buffering I could get. Our topic was justice and health care, and at the end of class I asked each student to write some question he or she thought the class had left unanswered. About a fourth of them wrote, roughly, "Why should people who make good money pay for health care for those who don't?" Two weeks later they handed in papers on a scarce-resource issue: if there is only one kidney available but two people need it, how should we choose? Virtually everyone argued that the single kidney should go to "Mrs. Benson" rather than to the unidentified accident victim: if it goes to Mrs. Benson we know it's going to someone worthy. The danger of rewarding someone undeserving loomed large for them; they would not take that risk. One wrote with stunning if unconscious cruelty that "the poor have worse outcomes, so treating the affluent is a better investment."

The ease with which they wrote such things, their unawareness of competing arguments and of a need to defend their own, exposed my failure. But the positions they took also said a lot about public opinion, particularly that of white suburban Michigan. I knew already what the day's election would bring—the polls were clear—and the student papers I was reading underscored the public mood.

So I welcomed the chance to get away, to drive across town for an IRB meeting. An IRB is an Institutional Review Board, responsible for protecting human subjects of scientific research. This particular board was part of the Michigan Department of Public Health. New to the committee, I had to get a sense of who the other members were, what the acronyms and jargon meant, what procedures people were used to. Handed a thick sheaf of government regulations, I appreciated, grudgingly, some of the resentment voters were taking into the booths that day. Afterward someone from the substance-abuse program lingered. Especially aware that day of hostility toward the government, I asked her whether she minded being called a "bureaucrat." No, she said; she was used to it, and believed that mid-level civil servants are advocates for the people in a way that no one else can be. My experiences with public health nurses and with community mental health workers inclined me to agree. Neither of us dreamed that the following spring hundreds of people would die in Oklahoma City, their only sin the fact that they worked for the government.

I drove back into town to meet with a hospital ethics committee. We were examining a draft of a policy on "Do Not Resuscitate" orders; it encouraged doctors to talk with their patients and tried to protect the wishes of incompetent patients even when their families have contrary wishes. This is a strong, mature hospital ethics committee, and I've learned a lot from them. That day, however, the committee was wrestling with the concept of futility, an idea that sounds perfectly clear until you try to pin it down. It had already generated a lot of academic writing, some helpful, some not. The committee was friendly but not entirely convinced when I insisted that the concept is too muddy to use, and that adding the adjective "medical" does not help. We struggled with language and promised to continue the conversation.

Finally home, I refused to watch the election results; tomorrow would be soon enough. Instead, I picked up an issue of the *Journal of Philosophy,* one of my few remaining efforts to keep up with mainstream philosophy. Confused in mid-page, I started again, and found that I had read the word "epistemology" as "epidemiology." Finally anchored in a familiar conceptual world, more abstract than anything I had dealt with all day, I read a discussion of contextualism that was a pleasure and a revelation. Charges that "We must contextualize!" are used widely, often vaguely and sometimes self-righteously throughout bioethics; here I found it distinguished from a related concept (coherentism) and defended with vigor and precision.[3]

The next morning, Wednesday, I finally tuned in the election results. At the time, they seemed from my point of view worse than I could have imagined. I tried to remember whether I felt as badly in 1972, or in 1980, but couldn't really recall. What bothered me most in what I heard was the anger and punitiveness. Political conservatism, as such, need have nothing to do with hatred of anyone, let alone of the poor; a fierce belief in individual liberty could be combined with an equally intense compassion. Principled attacks on taxation could be joined with vigorous calls for private charity. In Grand Rapids, for example, a city where conservative religion is powerful, a newspaper covered the death of a corporate executive by describing first, and at length, the man's civic spirit. His position in the business world was described three paragraphs down. Similarly, a single mother I know was urged by her Grand Rapids employer to keep in touch with her mother in India, and to call her from work so the company could pay for it. "When your children need you," he went on, "go home; we'll work around it." For many reasons, I don't in the end think private charity is enough, but it is important, admirable, and fully consistent with conservatism.

The public voice of conservatism in November 1994 was quite different. As political theorists point out, "safety net" welfare provisions isolate and stigmatize those who do not succeed, in a way that socialist and true welfare states do not. The language of that campaign showed exactly that stigmatization.

By Thursday the implications of the election had begun to unfold. For some (certainly not all) of what had happened I blamed the press, who seemed to lack much ability to reflect upon themselves. How welcome a journalism ethics movement would be. But the forces that gave birth to bioethics twenty-five years ago—technology forcing dramatically new choices, the civil rights movement, a set of scandals—have no analog in journalism. Its new technologies (computers, satellite dishes, the internet, and so on) do not force hard questions as kidney transplantation did, nor do they add weeks or years of misery to people's lives as ventilators and gastrostomies sometimes do. I cannot imagine, offhand, what social forces would lead the media to the sustained, public self-examination in which health-care professionals engage.

For that matter, I thought, I don't know what would lead universities to do so, either. My first project that morning was to develop a panel on academic ethics,[4] to be called "Devouring Our Young: The Mistreatment of Job Applicants, Part-Timers, and Junior Faculty." The panel addressed the way academics treat one another, rather than how we treat students, and that gave me pause. Bioethics, in contrast, has always focused primarily on how patients are treated, and rightly so. As the newly elected legislators promised to support higher education, I listened with mixed feelings. Like journalists, professors need to be considerably more self-critical. The most fundamental questions in higher education, I would argue, are about how many people should go to college, and why; perhaps we should close a significant number of colleges and universities, and put the money into primary and secondary education. Perhaps not. But there is no one positioned to raise the questions, and many to defend the status quo. Here as everywhere the questions that are actually asked are only a fraction of those that could be, and the difference is often a result of the distribution of social power.[5]

That point suggests the persistent criticisms of bioethics for its attention to individual cases and practitioners rather than to the social structures that distribute power. The criticism had never seemed fair to me. One need only recall the evolution of questions about kidney allocation, an issue with which the field began. A truly individualistic discussion would have remained at the level of individual choices: whether Mrs. Benson or the unidentified trauma victim deserved the single kidney. In fact, however, the discussion quickly moved on; the appropriateness of deciding on the grounds of social worth was soon questioned, and public policy almost everywhere came to forbid it (in theory, at least, and usually in practice). Questions of whether organs should be for sale, and of whether one's body should be considered property, soon developed. These are not questions about how one individual should solve one problem. All of us in the field help develop policy, from DNR policies in a hospital through fetal tissue guidelines in federal research.

One of my MSU colleagues, Leonard Fleck, has devoted his professional life to issues of justice. In 1994 his focus was the possibility of national health insurance. His work on such questions, moreover, has led him to develop a technique for working with audiences that he hopes will launch a more sophisticated public discourse on health policy. (One of my painful realizations in 1994 was that health care had played no part in anyone's election campaign that year. It was the first time I recognized that the Clinton health-care proposal of 1993 had truly and completely failed.) During the week I'm remembering, two of my colleagues planned a workshop with a major managed-care health system in Detroit. A few weeks later the *Hastings Center Report* would feature an article entitled "The Ethical Life of Health Care Organizations."[6] I concluded that the atomism of bioethics is vastly overestimated.

A colleague dropped by with brochures for the London course we teach each summer. (I had taken my turn the preceding summer.) An interdisciplinary class, its highlight is observation of English health-care professionals as they work. An early assigned reading, however, from the sociology of health, argues that health care doesn't make much difference; it is poverty and affluence that most affect morbidity and mortality.[7] In spite of much lower per capita spending on health care, and facilities that are often worn and shabby, health statistics in the United Kingdom are not very different from those in the United States. And in spite of quite an egalitarian health-care system, the health of the rich is far better than that of the poor.

With my summer in London vividly recalled, I began to write a review of David Hilfiker's *Not All of Us Are Saints*.[8] Again, I was writing for a regional newsletter, and I kept in mind Detroit physicians who might appreciate Dr. Hilfiker's "journey with the poor." In London's East End we had seen the poorest neighborhoods in the United Kingdom; yet I saw nothing like the devastation that exists in parts of Detroit. London poverty is nonetheless real, and I recognized its connection with sickness. One diabetic woman, for instance, wore tattered, narrow shoes, relics of a fashionable youth, but now a danger because they restricted the circulation of blood to her feet. Diabetics have a higher risk of gangrene and amputation, and her shoes increased her risk. In contrast, Hilfiker describes a diabetic who sleeps on a Washington street and gets frost-bitten; he is a homeless man who cannot wash his wounds, cannot refrigerate his insulin. The English woman and the American man are both in danger of gangrene and amputation, but his danger is more severe. Even if we had national health care, he would still be sleeping on the streets. His risk of amputation would still be greater than hers.

Suddenly all these things fit together, like pieces of a jigsaw puzzle. Yes, it matters whether we are asking the right questions; on the other hand, no, bioethics does not confine itself to the separate choices made by individuals. What bioethicists do, however, and however unintentionally, is to divert attention from the most serious health problem in the country, and one of the most serious moral problems: the grinding, killing poverty in which our underclass lives.

The next day, Friday, my day began with a meeting with Libby Bogdan-Lovis, an M.A. candidate in our Interdisciplinary Program in Health and Humanities. A childbirth educator, she believes that birth should ordinarily take place at home, and has data showing it to be safer. She wants to understand why birth in the United States constantly becomes more high-tech, more interventionist, more—she would say—brutal. Once again I remembered London, and a talk I heard toward the end of the course from Marjorie Tew. Then close to eighty, Ms. Tew was a statistician and, as she pointed out, about as neutrally positioned as possible on the issue of medicalized birth. She is not a doctor, not a midwife, and many years past her own childbearing. When she set out decades ago to find the data connecting improved perinatal rates with health care, she found none; she finally concluded that in childbirth, as in so much else, most of the credit for better outcomes should go not to medicine but to improved standards of living. (Nutrition is especially important. Rickets, for instance, deforms pelvises and makes normal delivery difficult.)

Thanks to Libby and others, my standard public presentation about what most in the field call "maternal-fetal conflict" now begins with the story of a laboring woman who locked herself into her hospital bathroom. She wanted a nonmedicalized birth, had negotiated for this plan with her obstetrician, and thought everyone concerned had agreed. In local labor and delivery suites, though, an intravenous line is routine, and the woman could find no other way to fend it off than to barricade herself. "Maternal-fetal conflict" is a bad name for a complex issue. What is ordinarily at issue is a disagreement *between a pregnant woman and a health-care professional,* both of whom care very much about the baby's welfare. The standard label, however, assumes a conflict *between a woman and her baby,* between the fetus's safety and her rights over her body. In the bioethics textbook I was using in 1994, only one article mentioned, and that in passing, a crucial case in which a woman whose baby was "certain to die" unless she had a cesarean section nevertheless delivered, vaginally, a perfectly healthy child. The author does not realize that this is a common outcome (in the few cases where the woman was able to avoid the intervention and where we know what happened). I know it because of having read work in medical anthropology—and from Libby, Marjorie Tew, and the British general practitioner with whom we worked in London.

Again, I realized that bioethicists tend to ask the wrong questions, or, more precise, not to ask some that are essential. It does matter whether a woman's rights over her body outweigh the claims of her fetus, the central issue in standard discussions of "maternal-fetal conflict," but it also matters that the predictions of doctors about harm to the fetus are unreliable. It matters, in part, because the moral conflict is less severe when it is less certain that the fetus is in danger. But it matters more deeply because the conventional account, omitting iatrogenic harm, not only exaggerates the role of medicine in saving lives but glosses over the unnecessary surgeries, episiotomies, hospital-based infections, and so on that result when birth is medicalized.

My point is not that medicine can do damage. Whether a woman gives birth at home or in the hospital, she and her baby are likely to be fine. My point is, rather, a broadening of my earlier thesis. Just as bioethics can divert attention from issues like poverty—death-dealing poverty—it can inflate the role of medicine in good health. We don't usually *say* that access to doctors is the most important factor in health and happiness, although on certain subjects, like prenatal care, we often assume it. The deeper point is that bioethics is newsworthy; we get calls from the media, invitations to speak, very often on questions of what doctors should do, and attention to us intensifies attention to health care, which comes to seem more important to well-being than it is.

On the other hand, all that attention, that ease of access to the media, means that we have a bully pulpit. I thought in 1994 that if I continued to give a voice to those who oppose medicalized childbirth, I would reach more people than Libby and her network could. I also thought that if bioethicists decided to pay more attention to poverty, we would have some chance of being heard.

My work week continued through Saturday and ended with what I still think of as the consult from hell. Many people were involved, most of them unpleasant, and they were filled with anger at one another. The issue was whether the parents should be allowed to choose certain risks for their child. Parents and attending physician were on one side, a specialist on the other, and the onlookers—a lawyer friend of the family and an official from the insurance company—remained inscrutable. I donned a red blazer, trying to remember whether or not it counted as a power suit, and almost immediately I realized that we should not have met at all without much more preparation. I could do nothing to facilitate respect, communication, or emotional processing. I acted like a judge and wrote an opinion like a judge's. Afterward I was anxious for days, worried about whether the child would live, knowing that what I wrote made his death more likely. Doctors live with this responsibility all the time, and in that week their burden became more real to me.

Philosophically, I found the case thought provoking. Standard terms like "best interests," "risks and benefits," did not seem completely adequate. I realized we need a vocabulary about possible futures and steps that start us on one path rather than another. The language of risk, strictly speaking, will do the job, but "risk" is a reifying and static word. We need something that suggests more explicitly a sequence of events, each stage having burdens and benefits, each choice making the following ones more or less likely. The specialist in this case described the parents as willing to risk their child's life rather than agree to minor surgery. But the boy will have nothing like a normal life span in any case, and the parents did not trust the only surgeon available. They preferred to chance an earlier death for their son than set out on what could, in retrospect, have been a futile path full of botched interventions.

The human elements were even more striking than the philosophical ones. One of my team members mused a few days later, "No one felt better when that consult was over." She was right; no one did. At the time, I had never considered "Participants felt much relieved" as a measure of an ethics consult's success. But on the day of that conversation I decided it was a necessary (though obviously not a sufficient) condition of having done well.[9]

At that point I realized how different my professional life had become. What I was doing in 1994 was not just different in degree from what I had done in 1984; it was different in kind. Ethics consultation, serving on IRBs, writing for practitioners rather than academics, and teaching medical students all demanded new knowledge (not of philosophy but of health care), drew upon new skills (especially interactional), and served different purposes (improving hospital policy, preventing the abuse of human subjects in research, relieving conflict over the care of a patient). I had joined a new field.

But Is It Philosophy?

After describing that week of November 7, the paper I wrote went on to raise questions that this book is an attempt to answer. At the symposium where I first presented the paper, in February 1995, it generated considerable controversy. Some questioned whether it counted as philosophy. My response was that I didn't particularly care, an impolitic remark during a conference honoring a book series called Medicine and Philosophy. I would not make the comment so casually today, but I would make it with years' more conviction. Part of the point of this book is to argue that bioethics is not a subset of philosophy, although it is well and uniquely fed by that discipline.

2 Bioethics as Something New

Many people still think of bioethics as a subset of philosophy. Others hold that at least clinical bioethics—the part devoted to decisions about patient care—should be thought of as a specialty within medicine and nursing. But most of the people with whom I spoke for this project were more open and in fact often somewhat puzzled about how to describe themselves and the field. Those who did feel clear were likely to identify with their original disciplinary orientation. One insisted, for instance, "I'm not an ethicist; I'm a philosopher. I don't give advice; I don't do consults."

But for my purposes, this speaker, who has written extensively within professional ethics, counts as an ethicist. I include under that rubric those who write, those who consult or are engaged in similar activities (task forces and commissions, for example), and those who create exemplary practices (for instance, model hospice units). "Giving advice" is an inadequate description of any of this, even of ethics consultation. The other interesting aspect of that speaker's remark, however, is the vigor with which it rejects the identity of ethicist. That rejection was not uncommon in my conversations. One person said, "I'm an anthropologist who does consults; I don't presume to have a deep understanding of ethics. . . . I won't misrepresent myself." Another reported, "I introduce myself by saying, 'I'm —— and I work in bioethics.' I don't know what 'bioethicist' means." The philosopher quoted earlier expects "medical ethics" to separate and become its own field, as I expect bioethics to do. In a later chapter I will argue that even now it is fruitful to think of the field as a practice, in the sense that Alasdair MacIntyre uses the term: as a coherent, complex set of activities, with its own distinctive goals and standards of excellence. His paradigmatic examples include chess, architecture, and portrait painting.[1] For the moment, however, I only want to explicate what so many of the people with whom I spoke implied: this activity is something new.

As background, we should remember that the fields from which people enter bioethics are themselves practices. This is obvious for the various fields in health care, but academics are often schooled to ignore that fact about their own disciplines. In the humanities—especially philosophy, history, and literature—one can all too easily think of the field as a set of disembodied texts. But these texts are written by individuals chosen according to explicit or implicit criteria (not just their record or their promise, but the schools they attended and the friends they have made) for a variety of motives and with assorted results. Postmodernists have pointed this out rather colorfully during the past decade, as have sociologists more quietly for the past century. But many fields remain un-self-conscious about their fuller selves and therefore un-self-critical about their customs. The customs, standards, goals, and skills of bioethics, nevertheless, are different from those of any of the contributing fields. The newness of bioethics, so startling to me at first, may make it easier to see the field fully, in all its practical as well as theoretical implications.

Not a Subset of Philosophy

In the 1970s, many thought of bioethics as the application of a philosophical understanding to a certain set of questions involving health care. Some philosophers still believe this. (I don't know of anyone outside philosophy who does.) In many ways the issue doesn't matter; what does matter is whether bioethics is worth doing, and whether it is being done well. The question does have relevance to academic turf battles: which department, for example, should be allowed to offer an undergraduate course in the field? In many schools bioethics is the only philosophy elective to attract substantial enrollments, and philosophy departments, often with very few majors, welcome and fight to keep these enrollments. The issue of the relationship between bioethics and philosophy matters in itself, however, when we try to define standards of excellence and decide how to train new members.

I see the relationship between philosophy and bioethics as something like that between theoretical science and the technological fields—between, say, physics and engineering. Practical fields are not just applications of theoretical ones: each level draws from, and contributes to, the other, and each has skills, standards, and goals irrelevant to the other. Lensmakers, for instance, do not just learn from physicists but also teach and equip them, give them crucial data, and provide tools that make new lines of inquiry possible. Biology, too, as a theoretical field, learns from doctors. It learns, for instance, that some procedures and medications work, sometimes long before scientists know why. Aspirin as a pain medication is one well-known example.[2] The same holds true between philosophy and health care, as Stephen Toulmin was one of the first to note.[3] Henry Richardson has explicated the relationship in more detail. Working toward a new theoretical model of the way general principles and particular cases interact, Richardson holds that we know what a norm really means only when we have decided what it means in practice.[4]

As I thought back over my week in November 1994, for instance, I remembered thinking that relatively little was at stake in whether Mrs. Bactri was given a gastrostomy. I now believe differently. Little is at stake if we think only of a few months added, or not added, to an already long life. A description in terms of respect for autonomy captures more: we could add months to her life only by failing to honor her convictions about it. A still more philosophically interesting description includes the way a death can change the meaning of a life. A deathbed conversion, for instance, changes the meaning of everything that came before; decades of life may now be seen as inauthentic, or, conversely, as preparatory for the final act. Something similar is true for deathbed victories: Mrs. Bactri's persistence vindicated the strength of her life-long convictions, and our willingness to respect them made her story a success rather than a defeat. Catherine Bactri's relationship to her mother also became a story of fidelity in the face of adversity.

In this example as in many others, birth and death, sickness and health, supply philosophers with important issues and insights.

Ethics in health care—thinking through questions and creating solutions—is not just a matter of "applying" moral philosophy. Each field nourishes the other. And each has distinct, if overlapping, purposes and methods.

The Example of Ethics Consultation

Many hospital ethics committees now offer the service of ethics consultation, and the word "service" is chosen deliberately. "Hospital ethics committees" are not disciplinary bodies. Instead, their purpose is to help hospitals understand and observe current ethical standards. An ethics consultation service addresses uncertainties and disagreements about particular cases. Those doing the consults share some sense of their purpose; after all, we talk with one another, write for one another, attend conferences together, and so on. Recently this consensus has been articulated by a task force of the American Society of Bioethics and Humanities (ASBH), the major professional organization in the field. Its guidebook includes a section entitled "Nature and Goals of Ethics Consultation."[5] I tend to use another formulation, outlined below, which I helped develop (coincidentally) about the same time with a task force at Sparrow Hospital in Lansing and later published in the *Journal of Clinical Ethics*.[6] The fact that the national and local versions closely resemble one another speaks to the consensus within the field.

Goal 1. "To promote an ethical resolution of the case at hand." This first goal concerns the immediate questions, those that led people to ask for consultation: In Mrs. Bactri's case, promoting an ethical resolution involved urging the doctor to respect her refusals. Work toward this kind of resolution demands moral analysis, in which philosophers are explicitly trained, but also a lot of what Haavi Morreim calls "gumshoe work,"[7] in which we are definitely not. "Gumshoe work" means finding out what's really going on. In this example, a consulting psychiatrist had said Mrs. Bactri was unable to make her own decisions. As it turned out, she was too deeply asleep to be wakened when he saw her. He had expressed no opinion about her waking abilities, which turned out to be normal. Gumshoe work also demands knowing the national consensus about such issues, and the state of the law. Neither consensus nor law is decisive, but one of our official responsibilities is making both available.

In general, this goal is less theoretical than it looks. When our task force spelled out what it meant in practice, we included advice like the following:

Determine whether the patient can speak for him or herself.
If the patient is able to do so, talk with him or her.
Urge that the patient or surrogate be fully informed about a range of possibilities, and about the likely consequences of each.
Help the patient and family sort out the practical and ethical implications of their choices.
Look for a pattern of choices in the patient's life.
Find out whether the patient has given someone durable power of attorney for health care.

Goal 2. "To establish comfortable and respectful communication among the parties involved." This second goal acknowledges that further decisions may have to be made later. Human beings are moral agents, responsible for thinking through their own principles and actions. They are also embodied, with various emotional needs and blinders. Furthermore, cases occur not as isolated events but as part of ongoing relationships. For all these reasons, a successful consult will help those involved work together as their story unfolds. When we met with Catherine Bactri, we tried to look ahead. Her mother was likely to develop aspiration pneumonia, for instance; would she want that treated with antibiotics? But we couldn't anticipate every possibility. Ideally we would have healed some of the tensions between doctor, patient, and family, so that they could deal with later decisions more comfortably. I'm not sure we accomplished this. And there was nothing in my graduate education that would have helped.

Goal 3. "To help those involved learn to work through ethical uncertainties and disagreements on their own." Here we look beyond the case at hand, to future situations involving the same kinds of issues. This goal does draw upon my philosophical training, since it requires that we write a clear rationale for our recommendations. But it's also true that many people untrained in philosophy do this and do it well. Furthermore, some of the questions that need to be answered have nothing to do with philosophy. (In particular: how does one best educate the health-care professionals involved? They may be hungry for theory, or indifferent to it; they may have wanted to think through an issue, or simply have hoped for legal protection.)

Goal 4. "To help the institution recognize ethical patterns that need attention." The pattern might be individual (a doctor who will never meet with families), or unit-wide (an ICU where patients are never allowed to refuse treatment), or hospital-wide (a confusing policy about "Do Not Resuscitate Orders"). Accomplishing this goal demands a practical perceptiveness, as well as a sense of the roots of the problem (individual psychology, an intrusive risk manager, inadequate staff training).

It is not just in ethics consultations that bioethics demands abilities far outside philosophy. One cannot write well in bioethics without knowing a great deal about how the world works, whether that is how the genome expresses itself, the financial realities of pharmaceutical companies, or the prognosis for very low birth weight babies. (It often seems sufficient to know these things rather superficially, a concern to which I will return in chapter 10, where I discuss interdisciplinary work.) Another example is that helping formulate codes, regulations, and policies requires an understanding of and the ability to work with those whose field is being regulated.

Not Just a New Version of Academic Life

So everyone entering bioethics, from whatever other field, must learn a great deal. They will need to learn the language, master a growing professional literature, become familiar with some history and paradigmatic cases, and usually acquire interpersonal and institutional skills. Because so much needs to be learned, it might be argued that what is developing is a new academic field. To some extent I agree, but I also believe the field does not quite fit even those wide boundaries.

There are many distinct practices within the academic world. Our vocabulary, and the institutional structures within higher education, can obscure this variety. The word "research" means quite different things to the laboratory scientist, an anthropologist in the field, and a philosopher in the library. The first may supervise a team of twenty people and a budget of several hundred thousand dollars, the second live with Stone Age people in the Kalahari Desert, the third immerse herself in the language of Aristotle. Each needs different abilities and has different experiences. Bioethics is at least as distinct.

One mark of the distinctness of bioethics is that traditional professorial duties become difficult to identify. The usual triad is teaching, research, and "service," a catchall phrase referring to everything from addressing a Kiwanis club through being an officer in a national professional organization. In the traditional academic setting, service is looked down upon, seen as a tempting trap for those whose time would be better spent publishing, or sometimes as a niche for those who can't do research. I don't particularly endorse that attitude; democracy within complex institutions demands a lot of time, and universities should share their resources with the communities that support them. On the other hand, the denigration of "service" has some truth to it; like crabgrass, these activities can absorb all the space available and yield little that is satisfying.

At first, a job in bioethics can seem like nonstop service. For one thing, little teaching seems to be expected: my medical school requires of each faculty member about thirty-five hours a year in the classroom, while liberal arts campuses demand five to ten times as much, from 180 to 360 hours a year. In addition, during one's first years, "teaching" may mean simply "precepting": meeting a group of eight to ten students, for whom the readings and assignments have been worked out by a course coordinator. The scope for creativity, and the amount of responsibility, is minimal. (It does remain possible to do the job badly, however.)

But these comparisons miss the point. "Teaching" in bioethics, especially within medical schools, refers to a substantially different set of activities than it does in traditional academic settings. Our formal assignments include not only precepting but also coordinating (designing the course that the preceptors are following) and guest-lecturing.[8] Informal teaching varies still more greatly, from person to person and place to place, but always includes talks and workshops for health-care professionals.

Each of these kinds of teaching presents distinctive challenges—practical ones and ultimately philosophical ones: Just what *is* teaching? Consider, for instance, precepting, an activity that I treated dismissively above: preceptors do not design their own course and do no lecturing. At MSU, we meet with students in small rooms with large tables, and one of my painful early memories involves standing for a moment and writing on the board, to make a point clear. The students responded with indifference, impatience, and disgust. This was not *my* show; it was theirs. I had stepped out of place. A preceptor who is not a doctor (and not a man) has a delicate role to fulfill: guide discussion, keep it on track, and offer no intellectual assistance unless it is requested. Eventually one learns, subliminally I guess, how to elicit those requests, and the enterprise becomes comfortable, but for me, as for many others, that learning took years. The need for such a delicate balance arises from various factors: from the conventions of "problem-based learning," a common format for instruction at Michigan State; from the attitude within medicine, already appropriated by students, that only physicians can speak with authority; and from the broader cultural assumption that there can be no such thing as expertise in ethics.

The result is that those two hours in a small room demand intense attention, listening, restraint, and care. Precepting is a different form of teaching, but not an easy one. And by its nature it presents a particular psychological challenge as well: it can be as punishing as a traditional classroom, since students can express their disinterest or displeasure as easily (indeed, more easily). But it cannot be quite as rewarding; the interaction is too brief and the preceptor's role not deep enough. Good precepting, therefore, often requires not only deftness but self-effacement, a sort of cheerful anonymous giving. Some similar differences hold for large-group lectures (where slides are virtually required, a scribe will take official notes to distribute to classmates who have paid for the service, and sometimes students sit in the back and read newspapers) and for curriculum design, which requires preparing perhaps forty pages of syllabus (objectives, discussion questions, assignments, forms for evaluating group participation, and so on) months in advance, and then "directing" fifteen autonomous preceptors. Even guest-lecturing is not much like teaching a single day of one's own course, since there is no history and no future: you make your point during that one hour, or you never make it.

Yes, those of us in medical schools teach. But what we call teaching is only kin, not twin, to what happens in a typical college classroom. Furthermore, teaching medical students requires specific practical skills and presents particular moral challenges. Some of our other activities might be considered teaching, too. Presentations for health-care professionals are clearly meant to be educational, and so is acting as a consultant to a hospital ethics committee. On the traditional university template, however, they count as service. The reason? Teaching is implicitly defined commercially and contractually. What separates "students" from others who learn is that students have paid for the privilege and will receive a credential for their money and effort. Teaching has to be defined differently in bioethics, without ignoring the fact that universities need tuition money in order to survive.

Just as in bioethics teaching shades into service, so does "research," especially because, in academia, research has come to mean work for publication. Someone working in bioethics writes for many different audiences and purposes. When the same point is made for several different professional journals—for instance, for radiologists, oncology nurses, and physical therapists—it can feel as if one is writing with a computer's "search and replace" function. Yet the point, say, about informed consent, may be new to each audience, even the bioethicists in that audience. The assumption in what is called research elsewhere, however, is that the point is new to everyone, that it is an advancement of knowledge. That's a fairly heavy requirement, and it can be hard to apply: what's known in one specialized field might be unknown in another, where it would be very useful. Or researchers can lose track of some point after it has been published. A former colleague who worked with the philosophy of Alfred North Whitehead once realized that the same small point about his metaphysics had been published three times, by three different authors, over a ten-year period. And although in some fields there is great difference in status between "original work" and "writing textbooks," in mathematics, prestigious departments will sometimes award a Ph.D. for a solid synthesis of work already done by others, but so scattered as to be inaccessible. In other words, the "original-not original" distinction is less clear than one might think, even in traditional academic fields.

All of which may seem like a plea: published work in bioethics should count as research. But that would be accepting too much, accepting academic convention as the standard for what we do. Instead, I think, we should be developing new standards, thinking more freshly about what it is we should be doing, and why. The uneasy fit between bioethics and existing academic paradigms, together with the practical, interpersonal, and institutional skills mentioned earlier, and the need for broad but not necessarily deep knowledge—all suggest that bioethics cannot be fully understood as a new academic discipline.

Nor a Form of Health Care

It's been natural for me to analyze bioethics by asking if it is a variety of academic life, perhaps one that has evolved into a new species. I came to the field, after all, through the academy, and because I'm a tenured professor within a university, my own life *is* a version of academic life. But bioethics is made up of many other kinds of people, and they would pose the question differently. A doctor or nurse, for instance, might wonder whether bioethics is, or should become, a specialty within health care. In one sense it already is; doctors and nurses do sometimes specialize in bioethics. On the other hand, the categorization is essentially inadequate. The most obvious reason is that the field's domain goes far beyond clinical medicine. But even if it did not, the field needs more than clinicians.

From the beginning there have been tensions between clinicians (doctors, nurses, and other health-care professionals) and others (usually academics). Many people believe that the next generation of bioethicists, at least clinical bioethicists, will for the most part be doctors. (In theory, nurses, too. But it is no surprise that nursing gets, for the most part, lip service only. I will turn later to this issue.) What right do academics have to pass judgment, clinicians say, when they cannot fully understand the clinical picture, when, above all, they do not bear the weight of clinical responsibility? Ethics courses in medical schools are sometimes introduced as a way to improve patient care and tend to be received better when those teaching them are physicians. One doctor who works as a paid ethics consultant believes only doctors can do consults well, because only they can catch medical oversights whose remedy will dissolve the problem (the patient is restored to competency, or even to health), and because only doctors can present themselves to troubled colleagues as people who understand and can help.

This doctor's position assumes that the goals of clinical bioethics are to judge and to offer assistance to physicians. Such a narrow construction is inadequate for many reasons. For one thing, many other kinds of professionals are involved in patient care and in the running of hospitals. And even when we are offering ethics consultation to physicians and in the kind of case where our goals include judgment and assistance, these will rarely exhaust our purposes. Mrs. Bactri's case included an element of judgment: essentially we backed the family against the doctor. We were also helpful to the doctor, since afterward he felt more comfortable about not placing a feeding tube. But the consultation felt more like a form of assistance to the family. After all, we reminded Catherine Bactri that she was free to change doctors, and we would have offered her help in doing so. Although it did not happen in this case, nurses can request ethics consults, too, and in such situations the consult helps compensate for power imbalances.

A practice is defined by its purposes, by the means it employs, the standards of excellence it develops, and the particular forms of failure it allows. Identifying these elements within bioethics is the central task of this book. One can approach the task from many different angles: What is implicit within the practice as it has evolved? What does our history suggest? What responsibility do laws, regulations, and court decisions give us? What does the public have a right to expect from us?[9] Part of my answer, one crystallized for me by the reflections of Judith Wilson Ross on ethics committees, is that we need to be agents of change.[10] And some kinds of change are more likely when the bioethicist is an outsider. I address this issue more fully in chapter 7, "Virtue in Bioethics: Choosing Projects Well." If I am right, then bioethics, even clinical bioethics, will always need to include a healthy number of what George Simmel called "strangers." And so bioethics cannot be simply a subset of medicine, nursing, or health care. It remains true, obviously, that clinicians and allied professionals (social workers, chaplains, and so on) are essential to the field.

Bioethics, then, needs a variety of skills and perspectives, and cannot be fully understood in terms of any of the occupations for which we now have names. I believe it can be usefully seen as an engagement in moral development, a mutual process involving ourselves and those we serve. Before turning to that argument, however, it may be useful to see the field through an allegorical lens.

3 Bioethics as a Territory An Allegory

The concept of *practice* is a useful analytic tool, and it will continue to provide a structure as my examination of bioethics unfolds. A backdrop of history, however, will be useful at this point, so that readers can understand the forces that have shaped the field of bioethics. Already, historians and leaders in the field have chronicled its history in a number of responsible, serious publications. I'm going to present the story in a lighter way, through an allegory, which will also serve to introduce the questions that later chapters treat in more depth. An allegory suggests there is more to the story than one is able to give, and allows for expansion and competing interpretations. This openness should be a useful starting point, and a complement, to the analysis I will present later.

The Land

Imagine a territory, something like the Oregon Territory, but as yet unclaimed by any national government and without indigenous people.[1] *For millennia there have been regular visitors from neighboring, ancient kingdoms, who have established base camps and cleared trails but do no mining or farming. For the most part, they use these camps as quiet places in which to polish goods they have brought from the kingdoms, which they then carry back with them.*

Before the fields of inquiry that we now call bioethics developed, people taught and wrote about ethics for doctors, for nurses, for therapists, for scientists. Their domain, however, was significantly different from what we now think of as bioethics. In most instances, these people were clinicians or scientists themselves, articulating norms into which they had been socialized, and exhorting their fellow professionals to follow them. Some came from religious traditions; they, too, were primarily repeating and explicating a set of norms shared by their readers, norms that changed slowly if at all.

Discovery

Suddenly this land attracts interest from outsiders. Some dramatic discovery draws attention to its possibilities: perhaps gold is found, and passages are forged over mountains heretofore impassible. Those who happen to be in the old camps when the newcomers arrive react variously; some welcome the interlopers, others try to fight them off. It becomes increasingly clear that the land holds more promise than has been noticed before, and that the ancient kingdoms might profit from its development.

The gold in this case was a set of new questions that offered intellectual interest, a chance to make a difference in the world, and the possibility of professional advancement. And for the first time these questions seemed to belong to the public domain. Sometimes professional barriers were lowered voluntarily and the newcomers welcomed, but at other times the walls had to be stormed. Medical scientists had not been adequately protecting the subjects of their research; doctors and hospitals did not want to decide by themselves who received scarce and life-saving kidney dialysis; in an era of civil and consumer rights, patients were growing dissatisfied with medical paternalism.

Immigration

A trickle, then a stream, of newcomers arrives, from many different homelands and for many different reasons. Some are missionaries, some homesteaders, a few are tourists, and a few are hustlers. Some are wandering aimlessly and stumble in. There are many hungry and adventurous young, and a few elders, hesitant, or thoughtful, or overbearing. Many have not been quite at home in their countries of origin.

The people who began to write, teach, speak and otherwise do the work of bioethics were not professionally trained in it. They came from religion, medicine, philosophy, law, nursing, the social sciences, and literature. Some came with a moral mission, a response to scandal or to some aspect of their own profession that disgusted them: "In residency once I was holding the bag—the CPR bag—for a demented woman in terminal pain. Nobody had a clue whether she would have wanted this or not. I swore I'd never be in that position again." Among those who had dealt with illness in their families, the quest could be moral in the broader sense, a desire for a deeper understanding of what had happened. "I was appalled at the way the desire for money influences the profession."

Some stumbled into the territory never having sought it out. Many of the people to whom I talked described the chance events that brought them to this new arena: They knew people who were taking on projects, or came across a grant opportunity, or "put themselves on the market" and ended up "sold" to medical ethics.

Someone started the center. . . . I got into it accidentally.

In graduate school I impressed my adviser as a good writer, so he hired me as a grant writer in the center.

Some were professionally homeless. People who had spent five or ten years of their lives earning a doctorate emerged into a new academic world, a fiercely competitive arena in which the losers were not simply undistinguished but unemployed. The humanities had entered an economic depression from which, thirty years later, no recovery is yet in sight: Many more people earned Ph.D.'s than could find positions in the field; those who did were given teaching loads that left no time for laundry, let alone for research and writing. By 1970 the departments who hired them were becoming hungry for enrollments; in the 1960s people had been willing to major in liberal arts, but as the Vietnam War wound down and the economy slowed, students turned to majors they thought would help them earn a living. The fortunate were hungry in a more traditional sense: young scholars who needed to publish, and hoped to build their careers around some new and exciting area of inquiry.

Others had a place to stay, but wouldn't call it home.

> I taught high school one year. After the first day I cried all night. I cried a lot that whole first year. I had to find something else.

> I hated [that university]. Yes, I was lucky to have a liberal arts job, but such conservative students and the community, too; I was literally spat upon for protesting Vietnam. The students were not interested in what I did.

That speaker gave up a tenured position in order to move into bioethics. Another gave up her chance at tenure: "In philosophy you're expected to be grateful just to have a job, any job, anywhere." But she, far from being grateful, was actively unhappy.[2] Someone else, from another of the academic humanities, felt not quite at home there intellectually or politically: "I was more interested in meta issues. I like to work with cognitive dissonance. And like the people I work with—so many 'old lefties' in the crowd—we want to retain some sense that we haven't entirely sold out."

Some were prospectors: academics, for instance, struck by the possibility of new questions and forms of work life.

> In my late 40s I took stock. . . . [W]hat resonated with me was activity that makes a difference. I continue to be dismayed by the narrowness of my philosophical colleagues.

> I'm a practical person. [I'm happiest doing something useful.]

> A lot of people go into philosophy in order to ignore the world. . . . In philosophy of physics we were allowed to look at the real world. But then I found out that philosophers of science didn't really know how science was done. . . . I liked logic, but didn't like doing philosophy of physics as if it *were* logic. [I guess for the same reason] I couldn't get interested in Judith Jarvis Thomson's way of doing ethics [through wildly fanciful analogies].[3]

Some who came from medicine were moved by intellectual hunger.

I was looking for some place to think. An antidote to the usual routine in medicine which is "Do, do, do—and think. Maybe."

I was numb from years of school and residency, and then suddenly I was out of the clinic, with time to do whatever I wanted. I began working in a facility for the chronically ill, ventilator-dependent, often somewhat demented; the families could be demanding. All the [other] doctors and nurses hated the work. These patients were the cast-off "successes" of our medical system. *What was this all about?* Suddenly I realized medicine could be interesting again.

Other physicians needed not stimulation but resolution. "I would have left medicine if I hadn't discovered bioethics. The medical part is easy: You look it up, and after seven years you should know how [to do almost everything]. But when I encountered what I came to call moral issues, I'd get a different answer from every senior physician. I needed some way to be sure that I had included all the significant factors when I made my decision."

Among the professionals who were young and came to the field when it was already partly developed, many report something like a conversion experience.

This was where my heart was.

I made my decision during a trip around the world. I wrote the letter in a tent under the stars, and I knew it was momentous; I made a copy for myself.

I just fell in love. I couldn't get enough.

Some were tourists. The mysterious and exotic world of medicine suddenly seemed open to select outsiders. A few have come to the field as hustlers. The field offers the opportunity to make a name or, if not quite a fortune, more money than in the usual academic life. (The contrast does not hold for medicine. A doctor mentioned that bioethics was one of the few specializations likely to reduce one's income.) Without credentials and standards, anyone could hang out a shingle.

Most, of course, arrived with mixed motives, somewhere on the spectrum between missionary and entrepreneur. They hoped to do well by doing good.

Settlement

Because the base camps are not adequate for the logging, mining, and farming that begins, the newcomers put up their own cabins; after a few years some of them make the new land their home. Some retain their original citizenship, but most maintain some kind of dual allegiance.

A few theologians and then philosophers took up the new questions. Some doctors and nurses did, too, in a different way than their forebears had done. These newcomers were not so much interested in transmitting accepted norms as in criticizing and changing them. At first these topics were only one of their many professional interests, and most retained their original disciplinary identity for a long time, unsure of whether this new "land" would sustain permanent settlement. Some, however, realized within a few years that they had developed a new professional identity, one that expanded or replaced earlier senses of self.

I'm always well grounded in anthropology; I haven't left it.

I don't write much for scholars any more; my audience is working professionals. My farewell to religious studies was gradual and easy.

I'd say I've abandoned social psychology; I haven't contributed any primary research for years.

First Years

Now that they form a group with similar interests, the newcomers band together for mutual support. Even the loners come into town occasionally for necessities and for company.

Settlements grow up around common projects: timbering, trading, farming, mining. At times there are temporary gatherings for the purpose of surveying the land, marking boundaries, and so on. Names for this new territory spring up in the way names do, pragmatically, accidentally, humorously. Some stick, but none seems quite right: the inhabitants are too different from one another, and the terrain too varied. The borderlands have less trouble, perhaps because they are smaller and more homogeneous.

People interested in questions of research and medical ethics talked with one another. Those teaching in undergraduate programs—especially in philosophy, religion, or science departments—developed courses for their own students. "Centers" and "programs" began to form within medical schools, with one or two core faculty and a number of part-time associates. Typically, but not always, these faculty developed courses for medical students. Task forces and commissions on the national level were less permanent, but often highly influential. Some centers formed independently of medical centers or universities, one or two people doing consultations about patient care, perhaps. Many different kinds of projects were undertaken: "white papers" for legislatures, conferences about the care of the disabled, changing professional codes of ethics. Many people began to sit on IRBs. Sometimes they did this work as volunteers; other times they made a significant amount of money for it. Hospital ethics committees grew up, and regional networks of ethics committees developed. Doctors and nurses developed pioneering ways of helping patients at the end of life, studied the results, and shared them with the professional and lay public.

The variety of activities is reflected in disagreement about an appropriate name: Bioethics? Medical ethics? Health care ethics? Medical humanities? Health and human values? My reason for choosing "bioethics" as the least inadequate is partly a result of being trained as a philosopher, where "ethics" refers to a broad field that encompasses all the moral dimensions of life.

Relationships

The settlements depend on one another. Some are near neighbors, others so distant they rarely interact. Some people own property in more than one settlement. Intermarriages take place. Trade and supply networks grow up. Personal feuds and territorial disputes flare; occasionally vigilante groups form. Some people become prominent, others are snubbed. Networks of privilege and prestige develop. Legends spring up.

A great variety in projects and in organizational structure characterized the field, hardly formed enough to be placed under one roof or name. Purposes ranged from the highly abstract to the thoroughly practical. People at one end of the spectrum did not always respect those at the other.

> I don't especially want to schmooze with doctors; if I've got any extra time, I'll go into my office, close the door, and think.

> The moral work of medical ethics is local. Most of what is published is pretty useless.

An infrastructure developed: journals, professional organizations, discussion groups on the internet. In much of this groundwork, however, participants could sense the tension between theory and practice. The Society for Health and Human Values (SHHV), one of the first organizations, was considered too abstract by some who were daily involved in clinical ethics, and the Society for Bioethics Consultation (SBC) was founded to meet their needs. But SHHV was disdained by others for lacking theoretical rigor, and that reaction resulted in the formation of the American Association of Bioethics (AAB), largely by and for philosophers. By 1998 the three associations had merged. Some humanists, those in literature, art, and history, feared that the new consortium would push them to the margins, or entirely off the page, and considered starting a separate organization. The tensions took other forms as well; some social scientists felt philosophers were missing the whole point (and sometimes that philosophy, by its nature, could do nothing else but miss the point); some doctors felt the field should belong to them and them alone; some philosophers condescended to everyone (including other philosophers of the wrong degree of abstractness or practicality).

Feuds and pecking orders proliferated. To caricature the scene: Philosophers dismissed religion. Social scientists disdained everyone, including themselves. Legal scholars valued it all, including themselves. Nurses were invisible. Georgetown graduates believed they owned the field. The young knew they were the future. Pioneers knew that they'd forgotten more than newcomers would ever know. Philosophers assumed no one else could do conceptual work; social scientists asserted that observation is all. Bioethicists who worked in hospitals called their work clinical and considered themselves the elite, while, in a ranking of philosophy programs by specialty, not one clinical program was listed under "Best in Bioethics."

But sometimes, somehow, these groups managed to work together. Often the relationship was one of trade, but there was an increasing amount of cooperation on common projects. Some philosophers may have felt in the beginning that their mission was to enlighten others, but no one can do this work very long without realizing that they have as much to learn as to teach, and finally that everyone is learning together as they face a common problem. A neonatologist might call me to see what the ethics literature says about tube feeding; I might spend time with him to see the practical implications of a decision not to tube-feed an infant; we might work together to write a paper on the subject. As is true in many arenas of collaboration, even what looked like an exchange (of information, say) was more likely to be some form of gift, placing parties in ongoing relationships of mutual benefit. Just as a variety of motives brought people into the field, so a variety of human relationships grew up.

Borderlands

Borderlands begin to be settled, land that shares some of the attractions of the Territory itself, but adjacent to different Kingdoms. The Borderlands vary, but without exception are smaller and more lightly settled than the Territory itself. Bor-

ders separating these new territories from one another and from the old Kingdoms remain rough: "We have everything this side of the river, they have always governed the land beyond the grasslands—but the river and the grasslands themselves? We both use them."

While bioethics was growing up, so were similar fields: business ethics; many forms of professional ethics, especially engineering and legal ethics; environmental ethics and the animal rights movement; and so on. Some feminist theory also is of the same kind; that is, it aims at understanding practical ways in which the world could be better. Motives for entering these fields varied as they did in bioethics. One person, whose family, in her childhood, had been involved in arguably immoral activities, believes that she was deliberately shown some of the details and some of the results. "They hoped I would remember. I was a kind of messenger to the future." Her work now is an attempt to make public, and stop, what once flourished in secrecy. Another person's motivation is more generic. He reports, "I love a good fight: one for a good cause and with a real chance of winning." The "borderland" fields differ in many ways. While many speak to specific professional groups (to engineers or city planners, for instance), some take all of us as their audience. An example is environmental ethics: the minds it hopes to change belong to no single occupation. Many issues belong to no one field, or overlap many: with the advent of managed care, for instance, those doing "medical" ethics become interested in business ethics. Problems of confidentiality and conflicts of interest arise in all the professions.

Just as boundaries between these new fields are permeable, so are boundaries between them and traditional practices (the "ancient kingdoms"). For my purposes I will roughly stipulate some perimeters, bearing in mind that these are not locations of honor but merely of description: On the academic border, those who simply teach undergraduate courses (bioethics, literature and medicine, and so on) are outside, although just barely. Those who teach graduate or medical students, or who write for publication about the issues, are inside. On the clinical border, those who "simply"

(!) try to practice as ethically and humanely as possible are outside the boundaries, but just barely; those who develop what I call innovative practice are within. By "innovative practice" I mean programs, like a number of hospice and palliative-care programs, intended to pioneer new kinds of health-care practice. I count them as ethics projects because they are not just instrumental toward already understood goals, as new forms of chemotherapy would be. The goals of these innovative programs are themselves newly created; they have been identified, argued for, and refined. Furthermore, the goals are intrinsically ethical, since they have to do with what it means to respect persons.

The Need for New Skills

Newcomers find that they need new skills: for starters, how to clear land and what to plant on it. Some things simply won't grow; others grow at first but quickly deplete the soil, or die during harsher summers. And how does one cook without familiar ingredients? Or stay warm and dry in this climate so unlike the one at home? How do people find their way without roads? Conversely, old skills are of little use: highway driving, supermarket shopping, staying safe in a big city. The newcomers who arrived here as adolescents have less adjusting to do. But they cannot just follow the patterns their elders have set; there are no elders, in that sense. Together everyone must construct a new form of life.

The first years in bioethics were filled with trial and error. What should we try to do, and how? What efforts will bear fruit, and is the fruit worth the effort? Within a few years, ethics courses were deeply rooted in medical schools, but even after decades not within veterinary schools. Hospital ethics committees became increasingly important, as accreditation agencies virtually demanded them, but their effectiveness was increasingly questioned. Regional networks of hospital ethics committees were important for a while, but later many of them withered. National commissions and other intersections with government came and went. Electronic conversation lists were established. At first there was great excitement at this new way to communicate, but within a few years many of the original contributors dropped off, overwhelmed by the volume or unhappy with the dominance of a few garrulous contributors. In spite of those problems, the networks became efficient ways to exchange certain kinds of information rapidly, and good places for newcomers to learn.

Those trying to decide what projects to attempt were usually working within unfamiliar institutions: hospital ethics committees, regional ethics networks, medical conferences. Even medical schools were alien to many. For a range of new endeavors, new skills must be developed: from the relatively trivial (using slides for presentations), through the obviously challenging (managing a volunteer organization), to the deep (helping clinicians reach decisions about patient care; making recommendations about national policy).

Some of the needed skills were, and are, moral. Virtue, in fact, is a kind of skill: a habit of doing the right thing at the right time in the right way. Habits of honesty and kindness, of fairness, of maintaining integrity in the face of pressure and temptation—one develops these through life in particular circumstances. Ways of being kind, for instance, differ as contexts differ; pressures on integrity vary as power structures vary. People in this new line of work occasionally find with dismay that they have compromised themselves, or fallen into dishonesty or unkindness, because the cues and challenges become harder to recognize. To take a single example, public speaking: What counts as manipulating an audience? When should one ask for a stipend, and how large should it be? Is there a name for the virtue of being able to look at audience feedback forms and dispassionately learn from them (rather than becoming defensive or distraught)? If one has little time, which invitations should be accepted, which declined? And how is virtue in these respects learned and taught?

MacIntyre characterizes virtues as qualities that are necessary for attaining the internal goods of a practice. (He also argues that courage, honesty, and justice are necessary for attaining the internal goods of any practice.) Aristotle thought of virtue as generally "lying in the middle," between excess and defect. Later I will look at the particular threats to integrity that this field presents, and explore the possibility that it requires particular virtues: discernment, humility, courage, honesty, and mutual respect.

The Young

The pioneers were not reared in this new form of life; they invented it. They carved space for it out of the wilderness. But, like primates who were not themselves nurtured as infants, some give no thought to supporting those who come after them. They cleared their own land; surely the newcomers should do the same? Others, however, form settlements that make everyone more productive and that easily integrate waves of new settlers as well as the next generation.

There is great disparity in the way newcomers to bioethics are treated. Some excellent things happen, both nationally and locally. The ASBH, for example, has an annual award for the best paper submitted by a graduate student, and gives the winner a steep discount in conference fees. In some graduate programs, students are proud and grateful for the help their distinguished teachers give them. Similarly, in some centers, veteran and new members meet regularly to discuss not only *what* they're working on but also *how, whether, and why* they're working together. On the other hand, there are programs where students are abandoned once they've graduated, if not before, and where junior colleagues receive only keys to their office and good wishes. I take these issues up at greater length later. My conclusion will be that where there's a problem, it is in part a result of the newness of the field. Each center is *sui generis,* so bioethics lacks the kind of standard path that has long been laid out for medical residents, newly hired nurses, and junior faculty in established fields. Centers that assume new members should carve out their own jobs, as senior members did, can put the young in impossible situations, since all the obvious slots may already be filled. And the veterans may not notice that a newcomer is struggling, busy as they are and misled by an assumption they have never articulated ("I cleared my own land . . ."). The young will, and indeed do, have a very hard time, waging an invisible and nameless struggle. Some will leave, and some will be forced out for "unsatisfactory performance."

Choosing Projects

After some years, the settlers become comfortable. In part they're self-sustaining—the climate is pleasant and certain crops grow well. Yet they depend for essential supplies on the older kingdoms, and they've learned how to keep those supplies coming. The need to please patrons, along with the habit of talking primarily with near neighbors, carries certain dangers. Will the community contribute to the world at large, or settle into cozy insularity?

Bioethics too became comfortable, and faced the temptations that come from insularity on the one hand, and from dependence on the other. The projects it takes up tend to be those that please the medical community, granting agencies, and general audiences. In later chapters I explore the moral challenges—especially to integrity and discernment—created by this need to please. There is also a problem of standards. New professional fora have developed—associations, conferences, and journals—but the work they generate can be lackluster. Journals of the "Medicine and X" variety can pale in comparison with the best work done in the original discipline. Since new journals address new questions, it is not surprising that they lack the depth and polish of the best work in older fields. But the contrast remains disquieting.

As years of effort accumulate, as opportunities to talk and learn from one another expand (a downtown grows up, in terms of my allegory), and if enough of us keep in touch with our disciplines of origin and with the outside world, all this is likely to change. In the meantime, I believe several things could help.

One useful step is an articulation, in MacIntyre's terms, of our defining purposes. Without clear boundaries, without even an uncontested name, what can we point to that unites the nurses, lawyers, chaplains, philosophers, doctors, and social scientists in this new line of work? We can say this much at least: each is trying to make health care, health policy, the biological sciences, and our shared understandings of them more deeply moral. And, as I argue throughout this book, we try to accomplish these things by contributing to moral development, broadly construed: growth in our own and our audiences' understanding of what a morally better situation is, along with the will and ability to bring it about. We have this aim not just as individuals but together, not just as a collection of like-minded individuals but as people whose professional lives depend on one another and constitute new entities: centers, journals, degree programs, commissions, and so on. Although I'd never argue that this field has a single, static, defining purpose, I believe it's helpful to organize our self-understanding around this notion of fostering moral growth, an idea that I explore further in chapter 6. Given some clarity about where we should direct our energies, we also need to recognize the forces that can lead us to misdirect them. I talk about this problem in chapter 8, "Goods We Want, Goods We Need."

Before talking about what unites us, however, I need to talk about something that separates us. We do not yet have an adequate language for our work.

The Language

The settlers arrive from many different homelands speaking different languages, and must find a way to communicate. A pidgin language develops, helpful for some transactions but unsatisfactory for others.

People who live and work together not only bond in community, they also fall victim to infighting. Within bioethics, certain disagreements develop and sometimes grow heated. Sometimes this is a matter of deep moral disagreement; physician-assisted suicide is one clear example. Other times, however, I believe the problem arises from people talking past one another. Too often there are no clear standards for resolving them, no agreement even on the meaning of the contested terms. In the next chapter I develop this part of the allegory. Bioethics would be a stronger field, I argue, if we better understood the kinds of language in which we work.

4 The Languages of Bioethics

We come to bioethics from diverse backgrounds: from nursing, law, religious studies, and many other professions. To some extent each of these "homelands" (in the allegory of the last chapter) has its own language, or at least a specialized vocabulary, and communicating between them can be a challenge. But we do it, and as a result a language peculiar to the field is growing up, a language of "autonomy," "informed consent," "ethics consults," "narrative ethics," and so on. In a certain sense all of us working seriously in bioethics are multilingual, at various times using a common shared vernacular (usually English), the specialized language of our profession (medicine, philosophy, and so on), and a language that exists only within bioethics. We might think of this third language as a kind of pidgin, an analogy that can be developed in interesting ways. But little attention has been paid to this multilingualism, and as result we fall into some significant traps. Not surprisingly, the young are particularly vulnerable. I will have more to say in a later chapter about the demands of interdisciplinary work; here I will focus simply on complexities introduced by language.

The Illusion of a Common Language

The first set of problems arises when people confuse the technical vocabulary of their training with the ordinary English from which it is drawn. In conversation with a friend recently, for instance, I suddenly realized that I use "imply" to express a logical relationship: for me, "X implies Y" means that Y follows logically from X.[1] For my friend, implication is weaker, a matter of hinting or suggesting. We went around in circles, as she said, "I didn't think he was *implying* Y, I thought he *meant* Y." I thought the same thing, but my language confused the issue. In bioethics the problem arises when a word taken from ordinary English has different technical meanings in different disciplines, and speakers are unaware that it does. The result is unnecessary confusion and sometimes dissension. Here are some examples.

Philosophers use "deductive" to refer to an argument whose conclusion follows with certainty from its premises, and "inductive" for arguments where the conclusion does not follow with certainty. Scientists more often use "deductive" to mean an argument from the general to the particular, and "inductive" for the reverse. The two distinctions often coincide, but by no means always.

For philosophers, "normative" statements say or imply that something is good or bad, right or wrong; in social science, the word is more likely to describe something taken to be standard, deviation from which is marked in some way.

"Empirical" has a much wider application within philosophy than within medicine. For this word, as for many others, it is important to look at what counts as a contrast. The boundaries change according to what the term is taken to exclude. For philosophers, empirical questions are those that must be settled by observation (understood broadly, to include the most sophisticated sort of scientific research and the most indirect kinds of observation). The point for us, roughly, is that empirical questions cannot be settled by reason alone. In medicine, "empirical findings" often means those that cannot be explained by theory; there is a sense of unsatisfactoriness about knowing something empirically without being able to explain it theoretically. A therapy whose success can be explained by biochemistry is more than empirically grounded; it is theoretically grounded. For philosophers its success remains an empirical fact, not a conceptual one. (We also know the distinction between empirical and conceptual is not water-tight.)

"Principle," even in philosophy, means different things in different contexts. It can mean an axiom, from which conclusions follow with necessity; it can mean a guideline or rule of thumb. To the scholastics, a principle was "that from which something flows."

The relativism which for social science is an almost sacred discovery is not the same relativism which is the bête noire of philosophy. Social science has established the depth and extent of cultural differences about morality; this is a *descriptive* relativism. Philosophers object to nihilistic conclusions sometimes drawn from these findings. The fact that people disagree about the rightness and wrongness of specific actions does not mean that they disagree at more abstract levels—hospitality, for instance, might be universally valued, even though its scope and expression vary. In any case, the fact that people now disagree does not prove that agreement is intrinsically impossible. Conversely, agreement does not establish truth. Even if the incest taboo, for instance, is universal in some meaningful sense, that would not establish that there is something really wrong with incest. The crucial point here is this: *arguing that something is right or wrong is a different conversation, demanding different kinds of reasoning, than showing that people agree or disagree about it.* The use of social science findings (not necessarily by social scientists themselves) to undercut moral conversation before it begins accounts for a great deal of tension between the two fields.

And, most crucial, "ethics" means something different from field to field. The philosophical field of that name is extremely broad, ranging from friendship in Aristotle's discussion of virtue to analyses of the structure of moral language. In the context of bioethics, philosophers are generally concerned with trying to understand the rightness or wrongness, the goodness or badness, of some action, policy, procedure, or approach. In this, we assume (as in other parts of philosophical ethics we might instead argue) that reason plays a legitimate and necessary role in this effort. Outside philosophy, however, "ethics" can mean description rather than prescription, and the study of it is empirical rather than semantic or moral analysis: "This is what these people find to be right or wrong, or to be within the moral realm or outside it; and this is probably why." The two sorts of inquiry can inform one another, but they are not the same.

Confusion about these linguistic matters may underlie my response to a claim I once heard a social scientist make: "From now on all ethics should be empirical." For a long time I could make no sense of this; it sounded like "all painting should be drawing." The two activities are not the same thing, and neither can be collapsed into the other. Granted, most great painting depends on draftsmanship, but there are fine drawings without paint and fine paintings in which there is no drawing. Similarly, excellent, abstract, ethical work does not necessarily depend on empirical work, and vice versa. But most important, work in bioethics requires a solid empirical underpinning—which is not always present. And if that is what the speaker meant, he has my complete agreement.

Many of the people with whom I talked made a moral project of tackling these language barriers. "I'm a translator"; "I often find myself translating"—that is, helping people from different backgrounds speak to one another. Another gave this example: "I taught an academic some lingo, told him to ask, 'Is she tubed?' so that he would seem like less of an outsider. Was I doing ethics? Yes, at least instrumentally." I agree. Helping us work with one another helps us be effective. Valuable energy can be wasted when we misunderstand one another.

The Comfort of One's Native Tongue

Still working loosely within the allegory of the last chapter, let me call the language one learns in graduate and professional training one's native tongue. The training is arduous and lengthy, but, once acquired, the language and worldview are second nature. They are comfortable and satisfying to work within, and it takes considerable effort to move beyond them.

One result, in the territory called bioethics, is that inhabitants can be invisible (or perhaps better, inaudible) to one another. Scholars in one field can overlook important work relevant to their own but done in a different discipline. Philosophers interested in narrative, for instance, tend to draw on other philosophers (e.g., Martha Nussbaum) and ignore the decades of work on narrative theory done in literature. We even tend to ignore the whole field called literature and medicine, a distinct "settlement" within the territory and with its own special product, the journal named after the movement.

A language is more than a vocabulary and a grammar; it is a way of structuring a form of life. It includes standards, at the simplest level for the application of words, but more deeply for what counts as valid and worthwhile activity. Any of us can fail to be self-critical about the standards we assume. A most painful, recent example was the attempt to bring philosophy and literary scholarship together within the pages of the *Journal of Medicine and Philosophy,* an attempt that ended in something close to bitterness. The philosophers found the work of the literary scholars not "philosophical" enough for the journal, a response that was difficult to swallow for those whose work had, after all, been invited.[2]

Conversely, we can incorporate the tools and language of others into our own projects and language, unaware that we have left behind the standards for their proper use. The result will be shoddy work. Two social scientists described with disgust a panel they had just heard: "Someone who 'solves' the problem of interrater reliability by having only one rater has no idea what reliability means."

The Language of Bioethics as a Pidgin or Creole

We shift back and forth between English and the technical languages of our professions, sometimes deftly, often clumsily. At times we also work in what might be called a pidgin, or a creole. Pidgins develop when groups without a common language interact with one another, and especially when they want to trade with one another. In the "initial stages of contact . . . a detailed exchange of ideas is not required and . . . a small vocabulary, drawn almost exclusively from one language, suffices."[3] "Pidgin languages by definition have no native speakers"; furthermore, like all languages, "they are social rather than individual solutions."[4] Creoles develop when a pidgin becomes a native tongue of a new generation, and languages exist in various stages between the two.[5] Of course bioethics is not literally either a pidgin or a creole—I am continuing my allegory rather than doing science—but the comparison can be illuminating.

Some may think that bioethics, instead, has created a jargon, and to some extent that is true. Jargon is obscure to outsiders but clear to insiders, for whom the words are precise and based on a common store of knowledge.[6] We use acronyms (DNR, PSDA, PAS), we refer to paradigmatic cases (Baby Doe, Baby K), and we have phrases created specifically for our work (ethics consult, human subjects, ethics committee). This forging of a new vocabulary happens in all fields. I am more interested, however, in cases in which language interferes with communication *among ourselves*. It is here that I think the pidgin analogy may be of use.

Others may think of the language of bioethics as a "dialect" of philosophy. At a bioethics conference, for instance, a social scientist remarked, "The discourse here is so foreign to me. At an anthropology conference I'm entirely at home, but not in philosophy." She, not on her own territory, assumed she must be on philosophical territory, that what felt foreign was a kind of dialect of philosophy. But I had been feeling equally off balance at the conference, and was struck by how different the setting was from, say, meetings of the American Philosophical Association. A crucial point I want to make, then, is that the language of bioethics is not, even metaphorically, a dialect of philosophy. Dialects are versions of a shared basic language. We might say, within my allegory, that endocrinologists and cardiac-care nurses speak different dialects of "medicine," and that postcolonialists and formalists speak different dialects of "literary criticism." Similarly, there are many different dialects of philosophy, and degrees of mutual comprehension. But there is a shared underlying stance, perhaps about the shape that philosophical inquiry takes, that allows us to recognize one another. Many people in philosophy do not recognize bioethics in these terms, and about many important aspects of it I believe they are right. (There is a subset of bioethics that does fit philosophical paradigms, however, and work within it could legitimately be considered a "dialect" of philosophy.)

What I want to show is that the language that has grown up in bioethics is enough like a pidgin that it is illuminating to press the comparison a bit. A basic point is that pidgins typically arise out of trading relationships. They are meant to be of use, and grow because and to the extent that they are. Much of the confusion to which I will point in a moment can be resolved if we turn to the practical questions at stake, putting aside grander and more abstract projects. Furthermore, it is always healthy to remember that each of us in some sense is trying to gain something, for ourselves or for those who sent us. In the preceding chapter I described the variety of motivations that bring people to the field.

More significant is the fact that pidgins are not what they may seem to be, simplified childlike versions of a mother tongue. The illusion that they are can, however, lubricate inherently difficult relationships. Real pidgins, like those that developed between the French fur traders and Native Americans, often arise in trading relationships between unequal parties; the language allows an interaction but keeps it from becoming close, and allows both to believe that the other is in some sense childlike. Conversation between, say, philosophers and physicians can foster the same combination of distance and condescension. One doctor, for instance, said of a philosopher bioethicist, "He could talk the [medical] lingo, but little things [he said] suggested he didn't have much clinical experience, didn't really know what it's like." A philosopher said of a physician bioethicist, "Well, the author's a doctor, so I guess I'm not surprised the conceptual stuff is weak."

Similar attitudes hold between other fields, reinforced by the nature of the language with which we communicate. Humanists tend to dismiss quantitative work as simple counting, the kind of thing a diligent child could do, and their dismissal is partly a result of not recognizing that a conclusion like "5% of X are Y" has to be understood in terms of research design, p values, and so on. The sentence may enter English removed from its evidential base—in fact, it probably will—and only scientists are likely to recognize what has been left behind. In similar fashion, clinicians can take an intensive bioethics course, learn "the principles," and become willing to dispense with nondoctor help. Some secular academics categorize religion as wish fulfillment and turn their back on the field of religious studies. All these instances of perceiving others as childlike, reinforced by the assumption that their language is a juvenile version of the original, are wasteful and lead to the denigration of important lines of inquiry.

If a pidgin is not an extremely simplified version of some mother tongue, what is it? A new entity. Its vocabulary may come from more than source language, and, upon transfer, words may change their meaning quite sharply. Its syntax is likely to come from a different source than the vocabulary. Examples in bioethics include the phrase "autonomy rights" when it is used casually to refer to all patient and family rights, even those, like parents' rights over infants, that have quite a different foundation. Not only vocabulary but the syntax changes, too, as in the phrase "consenting the patient." This phrase would be meaningless in any of the contributing languages, and reveals that theory has been transformed (or deformed) in practice. I should note that these particular examples come from bioethics as I have heard it spoken, not as I have seen it written.

The result of these complexities can be what is called a "Double Illusion." In the early stages of French-Indian communication, for instance, the French thought they were speaking a form of an Indian language, while the Indians thought it was a form of French.[7] In a similar way, people doing interdisciplinary work can think they are using the language of their "stranger" colleagues when, instead, they are speaking a new language that feels foreign to everyone.

Pidgins sometimes develop into creoles, and to an extent this is happening within bioethics; some of the new generation have acquired the language, possibly along with another discourse like law or medicine, as a native professional tongue. Creoles (and possibly pidgins?) can be more powerful in significant respects than the languages from which they are drawn.[8] To call the language of bioethics a pidgin or a creole is not to denigrate it. But it does call attention to certain features that may be limitations.

The analogy sheds light, for instance, on the fact that bioethics often seems limited in depth: the conversation has to be readily understood by everyone within it, no matter how disparate their training. Decades ago, K. Danner Clouser raised the question of "how 'deep'" biomedical ethics can be.[9] His own response was that, although the knowledge necessary for understanding the moral issues in medicine is vast, the moral theory required for them need not be. Clouser observed even then how frustrated philosophers could become with bioethics' rehashing of the same basic arguments, always drawing upon "an ordinary unsophisticated level of moral theory."[10] Doctors and nurses have a different form of the same problem. On internet discussion lines, for instance, doctors and nurses sometimes object that the discussion has become "too theoretical": not only difficult to follow but also impractical. What they want, instead, is more direct help with specific cases, and they don't want the practical and medical details abstracted away.

Some Examples of Bioethics as a Pidgin

I will begin with vocabulary. A number of significant terms within bioethics have a somewhat different meaning than they did in their "homelands": "autonomy," originally a technical term in moral and political philosophy; "informed consent," from the law; "culture," from anthropology; "workup," from medical practice; and "futility," from ordinary English. Sometimes the change is innocuous, sometimes helpful, but at other times it is confusing and even dangerous.

First, let us consider "autonomy." There has been a great deal of talk within bioethics of the "tyranny" of this concept. Yet when I was new to the field I was bewildered, hearing the word used in a variety of ways, none of which seemed right. Even when I understood that it was being used to mean something like the right to control one's own life, the talk of "tyranny" baffled me, since every day I saw patients virtually powerless within the medical system.

When I taught mainstream philosophy, I described autonomy as the ability, the corresponding responsibility, and finally the right of human beings to choose the principles according to which they live. (I also knew, of course, that such responsibilities and rights cannot simply be asserted, but must be argued for.) Autonomy is not a matter of having a choice; it is an inner stance toward whatever choices one has. Even a prisoner can be autonomous; even an emperor can fail to be (if he never rises above his whims or need for approval).

Almost nothing in bioethics engages these issues. Instead, rights over one's body (and sometimes a cluster of related rights, to privacy, to information, to decide for one's child, and so on) are called autonomy rights. Enormous effort, for instance, went into getting people to choose in advance what medical treatment they want at the end of life. Written "advance directives," many thought, would lessen the expensive and pointless interventions that many people say they dread and would ease the burden doctors feel as well.

Is this the most important ethical issue in health care? Arguably not. Are the proposed remedies—major efforts to get people to fill out forms about their wishes—likely to make a difference? Given the nature of institutions, probably not. Finally, are such advance directives appropriate for everyone, no matter what their background and situation? Certainly not. It's not only the case that many people don't make decisions without their families, something that is true across cultures; it is also, as I have learned from anthropologists, that some people, and peoples, believe they have no control over the future and therefore believe that putting dangerous possibilities into words increases the possibility of bad things happening. Forcing people to make decisions about future terrible possibilities can be cruel rather than respectful.[11]

When this discussion is conducted in terms of "autonomy," I am speechless and uncomprehending. Here, in this arena, I realize I'm not yet completely comfortable with the pidgin. But when the talk is of particulars, about problems, remedies, and worldviews, I have learned a great deal.

The problem with "autonomy" seems to me a case of the Double Illusion. Nonphilosophers understandably assume that "autonomy" as it is used in bioethics is a direct, unaltered borrowing from the language of philosophers. Most philosophers know it is not, but cannot say exactly what it is. Until at least this much is realized, the conversation is unlikely to proceed usefully.

Another instance of pidgin is the phrase "informed consent," whose home is the law, but which now peppers bioethics conversations in the most surprising ways. The phrase had a reasonably clear legal meaning: surgeons may not cut into patients unless they have consented, knowing what they are consenting to. A researcher may not risk injuring human subjects unless they have freely agreed to run the risk. Essentially "informed consent" referred to a right to say no. But the phrase now is used, informally at least, for a vast array of patients' rights. Alex Capron offers an analysis of why that is so: "The [legal] doctrine of informed consent—in reality, many different doctrines varying in their particulars by jurisdiction—has

. . . helped to reshape the physician-patient relationship in myriad ways, many of them unanticipated. . . . Of course, the effects of the legal doctrine cannot be separated from the effects [of] other forces. . . . [But] the basic theme that has been pounded home . . . [is] that physicians should engage in a dialogue with their patients before intervening in their lives."[12] This is a case where the use of the pidgin may be surprising (if one knows the term's roots) but seems to do little damage. It can, however, keep other morally relevant factors in the background, and so contribute to the lack of depth noted earlier.

A third example of a word that has entered the pidgin of bioethics and lost much of its original sophistication is the word "culture." (To be fair, it has entered ordinary English in much the same way.) Anthropologist Susan Long studied attitudes toward death among Japanese and among Americans. She found great intragroup variation (Japanese do not all think alike on the question, nor do Americans), and a significant amount of overlap between the two groups. She concluded:

> I am concerned about the way bioethicists and medical practitioners, even those who are trying to be sensitive, utilize the concept of culture. I begin my work as an anthropologist skeptical of ready-made categories and definitions. These are often derived from statements of religious leaders and from religious texts and represent the way *all* people feel about as much as the statements of the Pope reflect American Catholics' attitudes toward birth control. . . . [S]o although I have cautiously made some generalizations, I do not believe that they are predictive. It is all too easy to respond to an African American family's emotional expressions of grief by asking, as a doctor did me, for information on African Americans' beliefs about death, or to say, as a social worker in the hospital did in a pediatric case involving a Filipino mother, that the parents' attitude must be an "Asian" response to the crisis of their child's serious illness. . . . Variation within societies may be as great as variation between. *Culture is not a demographic category, but is the living of daily life.*[13] [Emphasis added.]

Here we have a pidgin that is adequate for simple sorts of exchanges; it reminds people that others can be quite different and encourages respect in spite of that. When it is used more broadly, however, and a more sophisticated exchange is needed, the word "culture," as it has been understood, does damage.

Still another example is the concept of an ethics "workup." Ethicists have borrowed this term from clinical medicine along with a variety of other words including "differential," "consult," "conference," "clinical," and "fellowship." James Drane, for instance, presents an "ethical workup" that breaks decisions into four phases: expository, rational, volitional, and public.[14] David Thomasma has created an "ethical workup" that has a series of steps, from describing medical facts and moral values through determining the principles in conflict to choosing the course of action to be followed.[15] This form of doing ethics is not the work of specialists; it is not comparable to, say, a neurology workup. If anything, it is the opposite, a simplified form for those who only occasionally must resolve moral problems. A workup within the medical setting is a systematic attempt to test various hypotheses about an illness. Although ethical workups are systematic, they do not involve any form of testing hypotheses. Ethical workups are only vaguely like medical workups. But they can be useful guides to moral thinking as well as a means of communication. In this way, then, the concept of an ethical workup is an example of a pidgin that functions well in the bioethics community and has not been overextended.

Finally, when bioethicists borrowed "futility" from ordinary English and adapted it for their own uses, they gave it a distinct meaning. In bioethics, the word refers to situations where *the health-care team believe* that an intervention is not worth doing (either because the chance of success is very low, or because what counts as success is not worth seeking) and the patient or family disagree. The literature on the topic does not include situations where the position of family and clinician are reversed; what is written on this issue falls under the rubric of the right to refuse care. Furthermore, "futility" cases are almost never those of real physiological futility, like the use of antibiotics for viral infections or laetrile for cancer.

Many bioethicists have pointed out that "futility" discussions are almost always value-laden, rarely if ever simply disagreements about facts. But the use of a term that in ordinary English can mean simply "this won't work" obscures the real nature of the debate. Furthermore, in English the word ordinarily refers to a sad and frustrating situation, but never (by itself) to an intense disagreement, nor does it suggest in any way the sense of compromise and complicity that haunts clinicians who feel forced to provide what they believe is futile care. "Futility" within bioethics evokes many more moral issues than it does in ordinary English. Once again we have a word that works well at an elementary level, that is, as a shorthand reference to a common and important set of issues. But it is not helpful, in fact it interferes, with attempts to discuss and resolve the issues.

Beyond Vocabulary

Besides vocabulary, every language has a syntax: formal rules of word order and morphology that carry meaning. (Morphology refers to changes in the spelling of the word in order to indicate number, person, case, tense, and so on.) In addition, most languages have a literature, oral or written, that includes distinctive genres. When the rules and forms of one language are mixed with those of another, the results can be unsatisfactory. A physician who also held a doctorate in one of the humanities said to me with some heat, "All my publications are in non-medical journals because I hate the way medical articles are written. Being forbidden a substantive endnote is a moral outrage." He went on to describe the language of medical journals as simplistic, lacking depth and nuance. Similarly, a philosopher found it "self-evident that the ethics pieces in medical journals are thin."

The thinness is an inevitable result of the "translation" into pidgin—of compressing what would be a thirty-page article in a philosophy journal into three pages in a medical journal, omitting the footnotes that would normally amplify and modify, fitting one kind of analysis (moral) into the framework of another (scientific). As a physician ethicist put it, "I was asked to write a 'structured abstract'—purpose, method, results, conclusion, discussion—that just doesn't work for ethics." In spite of these difficulties, ethics articles in medical journals are essential to bioethics. My analysis is meant to help us understand the problem, and not just rail against it.

We might, then, look at certain conventions of professional writing (such as the "structured abstracts" of medical journals) as if they were rules of grammar, ways of arranging words and symbols that are essential if they are to carry meaning. Take documentation, for instance. Most medical journals allow only a certain kind of footnote or endnote: it should contain citations to other works, and that alone. The note may not include reflections, qualifications, amplifications, or worries. To some extent this rule expresses the assumptions of what Kuhn called "normal science": the assumption that progress is a matter of adding to or subtracting from the store of accepted facts. A uniform logical structure is also assumed: the article shows that something is (likely to be) true; the methods and findings of the study reported are evidence for that; the footnotes are usually evidence from other studies for the truth of sentences in the article. Furthermore, science writing demands a crisp sense of what counts as the core of the work: all of that core, and nothing else, should appear as text. (It is instructive to read the original letter to the editor of *Nature* reporting the cloning of Dolly, and compare it with discussions elsewhere.[16] The prose is perfectly dry, understated, and factual, a description of one particular event, with no hint of what its implications might be. An uninformed reader would never guess that the event would reach the headlines of every major newspaper within a day.)

Scientific publication also has a clear set of genres: research reports are one thing, editorials another, meta-analyses still another. Finally, standard scientific research reports follow a strict format: problem, methods, findings, discussion. These conventions make for the efficient transmission of results, and provide tools the reader can use to evaluate them.

The humanities differ from science and from one another. Sometimes progress is additive (some kinds of history and ethnography, for instance). But progress can be of quite different kinds: new questions, new frameworks, and new concepts, for example, are highly valued. I once said to a mathematician friend that the best work in philosophy is true, original, and significant, but that managing all three was so difficult that accomplishing any two should count as a great success. He replied that surely truth is more important than the other two? I had to say, not necessarily. A fruitful mistake can be more valuable than a trivial truth. I imagine this ranking is true in most of the humanities but rarely in scientific writing.

In sum, publications in the various fields amount to records of fairly different kinds of endeavors, and as a result use different conventions for reporting these endeavors. In history, an essential question is whether the sources adequately support the conclusion reached; this is close to the scientific format, but not identical, since the footnotes will often extend the discussion, reflect on related items in the sources, and so forth. Philosophers also use footnotes to reflect on implications, compare positions, and so on, but virtually never use footnotes to establish that something is true. When footnotes refer to other philosophical writing, their purpose is to give credit for the ideas that stimulated the writer's thinking, or to locate the conversation within which this piece is placed. The footnotes can help a newcomer learn the territory, but they do not serve as evidence for the claims being made in the text.

The arrangement of words into footnotes, then, a kind of grammar, conveys a quite different meaning in different fields. And bioethical writing can marry the conventions of different fields in startling ways. Sometimes clinicians, for instance, simply present as true the conclusion of some philosophic analysis, never discussing what kind of argument supported the conclusion or how controversial it is. This entirely innocent misuse of philosophy shortchanges readers, who may have no idea of what they are missing.

A second result of joining humanistic content and scientific form is writing that is stripped of its power, rhetorical as well as epistemological. One measure of good anthropology, for instance, is the clarity with which the voice of the author can be heard. Science tries to erase that voice. Philosophy is a matter of the most explicit argumentation or conceptual analysis, work that cannot be summarized in the way data can, so the typical philosophical article is many times as long as a standard scientific piece. When the philosophy is cut by three-fourths, changed to passive voice and third person, and stripped of the qualifications that could make it precise, something new has been created. Read as either science or philosophy, it will fail. Thinking of it as work in a pidgin language, created to allow communication across cultures for mutual gain, may help us evaluate it as something other than bad philosophy or bad science. It is a new kind of writing with new goals.

We need standards for this new kind of writing, new standards not simply borrowed from any of the contributing fields. These will evolve as the language moves away from pidgin and toward a creole, which by definition has many more resources. Some of the current works that draw upon, say, Aristotle's treatment of virtue, or Carol Gilligan's moral psychology, or the history of casuistry are wonderful pieces, bringing disciplinary work to broader audiences and helping to create a shared intellectual world. But too much is still written in a pidgin: the fundamental terms are wrenched away from their theoretical homes, oversimplified and distorted. The result can be futile controversy over cartoon-like theory.

Interim Solutions: Awareness, Compromise, and Plain English

The issues that arise from what I call syntax—issues of the proper form for bioethics writing, and the appropriate use of documentation—become urgent for graduate students and young faculty. Those colleagues who judge work by apprentices need to be self-conscious about the standards they are using and be willing to compromise about them, where "compromise" does not mean lowering one's standards but changing them. On an interdisciplinary thesis committee, for instance, a scientist should be willing to accept first-person writing and substantive footnotes; a philosopher should recognize that the thesis is being written not for philosophers but for a wider audience, and allow some analyses to be abbreviated.

For questions of vocabulary, although all the languages of bioethics have advantages and pitfalls, as far as I know the only pitfall of plain English is that it takes more time. I spoke earlier about some ways to make the "autonomy" debate more manageable. Something similar could be done in many situations. Instead of invoking "autonomy rights," we could remind one another that patients have a right to something in particular; instead of "informed consent," we could say "patients have a right to understand the range of options, to refuse any of them, and to be part of the process of choosing among the rest." These phrases are clearer and less controversial than the slogans of bioethics. No one today would deny, for instance, that patients ordinarily have the right to reject what doctors suggest. Disagreements among bioethicists are narrower, about the relative importance of the right to refuse, and about what in practice should be done to honor it. Habitual use of plain English would prevent some of the odd locutions and muddy disputes that occur in the pidgin.

On the other hand, the English phrases lose something because of their practicality and particularity. The pidgin "autonomy rights," for instance, organizes a complex bundle of profound if imprecise moral convictions about what it means to be a person; using the term perpetuates these convictions and makes us at least question the rights of, say, husbands over wives, or elders over young. The scope of the phrase is both a strength, in the questions it insistently raises, and a weakness, because some elements in the moral convictions it bundles together deserve to be distinct and controversial.

Perhaps a creole will develop that carries the moral weight of the pidgin and is capable of the precision found in English and in disciplinary vocabularies. Until that happens, writers in bioethics need to choose from three kinds of language, conscious of the strength and limitations of each: the language of their home discipline, usually a vehicle for precise and efficient communication among peers; the pidgin of bioethics, sometimes the best or only way to discuss basic practical points while keeping a moral dimension in the conversation; and—for now the best, if the hardest, choice—plain, careful English. For my part, I rarely use either "autonomy" or "futility." I use "informed consent" only in its original sense, but the broader uses I hear do not seem to cause any problems. "Culture" now makes me stop and listen more carefully, as I have recognized (with help from social scientists) that it can be used to reinforce stereotypes and exclusion.

A few practical projects would help. Both philosophers and social scientists are interested in the way a word or phrase actually functions. I tried to capture this function earlier for the word "futility," and quoted Susan Long with a similar purpose on the word "culture." Philosophers are likely to investigate a term's meaning by analyzing language, social scientists by observing real-life situations. There is considerable overlap in the points to which they pay attention: both are interested, for instance, in implied contrasts (the "as opposed to what?" question) and about the practical implications of using one word rather than another. If the controversy about "autonomy" continues, or about similarly central terms, it would be helpful to know more about how the word functions in day-to-day, practical bioethics. My own sense about the word "autonomy" is that, first, it often functions to shift power away from physicians and toward the family (and occasionally toward the patient). Second, I believe it sometimes serves to diffuse the crushing sense of responsibility clinicians can feel. I hear, implicitly, from physicians, "Won't someone please say that it's all right for me to do this?" Even as I write, however, I realize there is much more to this picture. My real point is that we need to stop fighting about the supposed ideological or disciplinary roots of contested concepts and look at the role they play in real decision making. We might be surprised.

About the future languages of bioethics I will not speculate. I do believe that communication among ourselves will be more fruitful and less exhausting if we become clearer about the languages from which we choose. Each is suitable for certain purposes, none for all. If we see these advantages and shortcomings more clearly, perhaps we will be able to use them more intentionally.

5 Bioethics as a Practice

Why is it important to understand the languages we use? What unites us as bioethicists, despite our diversity of background, projects, and motivations? I believe it can be useful to think of bioethics as a practice, as the term is used by Alasdair MacIntyre.

A practice is a coherent and complex set of activities, socially constructed. It has distinctive goals and standards of excellence that help make the practice what it is, and that cannot be fully understood apart from it. Chess is a practice, checkers is not; architecture is, but bricklaying is not. A brilliant opening gambit in chess is a good example of a kind of excellence that cannot be understood if one does not understand the game; checkmate is a goal that cannot be understood without knowing what chess is about. These defining goals and standards develop through time; portrait painting, for instance, once aimed at symbolizing the subject (by means of objects arranged around him, say, and by his attire) rather than at capturing a good likeness; what counts as a good likeness also evolved, so that, beyond a recognizable image of the subject, what is wanted now is an expression of the inner person. Each practice is lived out in a specific way: the lives of a physicist, a rancher, and a portrait painter differ significantly from one another.[1]

MacIntyre introduces the concept of a practice in order to talk about virtues. He argues that all practices, no matter what they are, require courage, justice, and honesty, but also that different forms of life, with their distinct practices, will differ somewhat in the virtues they require. Different eras of our own past, for instance, found different attributes to be morally praiseworthy. Homeric heroes needed different strengths than did Jane Austen's heroines. Benjamin Franklin praised thrift and hard work; Aristotle did not. Although these are clearly expressions of cultural difference, that fact does not end the moral analysis. Franklin's position *may* be simply ideology or moral propaganda, but it could also be argued that within capitalism thrift and industry are truly moral goods, because without them one is not contributing fairly to a system from which one will benefit. (The point would have to be limited, since capitalism needs criticism as well as support.) In any case, the distance between Franklin and the Greeks nicely illustrates the idea that different ways of life present different moral demands. My reason for examining bioethics as a practice is to look at its particular moral requirements.

Bioethics may not be sufficiently developed to count as a full-fledged practice; perhaps it resembles bricklaying more than architecture. I will not tackle that question. For my purposes the field is close enough to being a practice to make MacIntyre's treatment illuminating. My major point is that, practice or near-practice, bioethics should be evaluated as a practical part of the world. It is not just a body of academic writing, and the practice of it cannot be evaluated adequately on solely academic grounds.

First, then, let me flesh out my sense that bioethics constitutes at least something *like* a practice. Its complexity is obvious. Our professional lives involve not just reading, thinking, and putting words on paper but many different kinds of learning and sharing. My own week in November 1994 was varied enough, but when we add what others do (and what I myself do at other times), the complexity becomes still clearer. Working on an animal-care committee, writing grant proposals, teaching medical students and undergraduates, writing for academic journals and informal newsletters, consulting on patient cases, presenting hospital Grand Rounds, helping federal agencies make policy, offering a listening ear to clinicians, organizing conferences, running networks of ethics committees, working with legislatures, modeling new ways of providing health care, speaking to the press, and so on and so on: this is a complicated life.

Its coherence is almost as obvious. The shape of each life in bioethics is somewhat different. Yet I think most of us find certain currents flowing through our activities: what we learn from practicing professionals, for instance, we share with students; challenges offered by students shape what we write. And so on, each activity informing the others. Later I will suggest some ways of conceptualizing these unifying currents, but now I will simply offer an example of the way the activities fit together by pointing to the issue called "physician-assisted suicide." Throughout the 1990s bioethicists responded to Jack Kevorkian, the Michigan physician who "helped patients die" and sought the limelight for what he was doing. The wave of deaths elicited a lot of bioethics activity: we taught, spoke, wrote for one another, developed *amicus* briefs for the Supreme Court, examined what was happening in the Netherlands (where euthanasia is accepted for certain situations), developed better forms of palliative care, and looked more closely at the way people die here. Legal argument, social science research, exemplary practice, philosophical exploration, comments to the press, workshops for clinicians: these are very different activities, but each was part of a coherent whole.

On the other hand, the coherence should not be exaggerated. Practices do not have Platonic essences, fixed and immutable. They are human enterprises, changing through time, shaped by the circumstances in which they grow. (Bioethical interest in "managed care" exploded during the 1990s as those structures came to dominate health care.) The future boundaries of bioethics are unpredictable; the changes will be to some extent organic, and to some extent a matter of historical accident. If the changes are sufficiently radical, bioethics might not remain the appropriate label. To count as bioethics, I would argue, the activity must deal with moral questions that arise about health care, health policy, and the life sciences.

Finally bioethics, like every other practice, is social. It cannot be learned well in isolation, because it is a matter of interacting with people and institutions as well as with ideas.

Defining What We Are: A History of Views

MacIntyre pointed out that each practice has internal goals, understandable only in terms of the whole and partly constitutive of it. What are the bioethical equivalents of, for instance, checkmate?

This is a question about purposes and about distinctive ways of reaching them. The two are intertwined: the point of a race is not just getting to the finish line, but doing so in the way specified by the sport, whether that means swimming, skating, or running. One central goal of bioethics is to expand our understanding of the moral dimensions of health care and biological research, at the public as well as the individual level, and to help translate these ideals into practice. But we are out not only to understand but to make a difference; what most distinguishes bioethics from, say, academic philosophy is its relationship to the lived world. These purposes shape one another. Definitions of death, for instance, are not just clarifications but decisions, made not just in the light of physiological facts and conceptual coherence, but also in the light of what can be accepted by clinicians and patients and even what the implications for organ donation are.

The nature of our *intellectual* tools has been hotly disputed for years: should we be using principles, cases, narratives, all the above, or something else entirely? Our practical tools, with the single exception of ethics consultation within hospitals, have gotten little attention.

In the beginning, the academics who took up practical ethics thought of it as "applied ethics," solving ethical problems by applying ethical theory, especially general principles like those of utilitarianism ("Do whatever will produce the most happiness and the least suffering in the long run") or Kant's deontology ("Do only what you could consistently will that everyone do," "Treat human beings as ends in themselves, never merely as means"). The idea was to add to these principles whatever empirical information was available and then deduce an answer. For instance, keeping utilitarianism in mind, people asked what happiness and suffering would result if we allotted a scarce life-saving technology (like the original kidney dialysis machines) according to criteria of social worth, choosing, say, the mother of six over the town drunk. Questions like these are enlightening. (Allocating according to social worth, to continue the example, would foster a humiliating anxiety: "If I get sick, will be I be judged worthy of help?" This would be a significant kind of suffering.) Only a convinced utilitarian would take this line of inquiry as final, but almost everyone thinks that questions about consequences, about happiness and suffering, are important parts of ethical inquiry.

But even twenty-five years ago, the "apply the principles" model did not capture what practical ethics was like, not even if we confine ourselves to the activities of philosophers. For one thing, few people are completely convinced utilitarians or Kantians; even those who are do not find their principles leading to simple, clear answers. In addition, the field has always used a variety of kinds of inquiry. Appeals to consistency, for instance, are commonplace in ethical analysis; moral agents require it of themselves. These appeals take the familiar forms of generalizability (What if everyone did that?), reversibility (Would you want that done to you?), and analogy (as in Judith Jarvis Thomson's famous abortion-and-the-violinist article).[2] Moreover, basic concepts need, and from the beginning were getting, analysis: just what is privacy, for instance, and why does it matter?

Furthermore, the most important work of practical ethics done in the twentieth century, John Rawls's *A Theory of Justice,* explicitly endorsed what he called "reflective equilibrium": a moving back and forth between attractive general principles, on the one hand, and possibly conflicting convictions about particular cases on the other, checking and revising each level of conviction in the light of the other. *A Theory of Justice* was published in 1971, when bioethics was in its infancy.

So practical ethics was never simply an application of principles. Yet the label "applied ethics" stuck for a long time. Bioethics had its own specific version of this story: a misleading label took on a life of its own and remains today a target of attack (in my opinion, pointless attack). The story began in 1978 when Tom Beauchamp and James Childress published *Principles of Biomedical Ethics;* now in its fifth edition, it remains a leading textbook in the field. That landmark first edition aimed at bringing coherence to a field where discussion seemed scattered among many separate problems. As a result "the moral judgments involved in one dilemma [appeared] to be unconnected to the moral judgments in others." The authors argued that a small set of principles underlay all the particular dilemmas, both giving rise to them and being useful in resolving them. Beauchamp and Childress went on to list four principles eventually known (usually derisively) as "the Georgetown mantra." (Both men then taught at Georgetown University.) The principles were autonomy (the right of competent patients to make their own decisions about health care); non-maleficence (the duty of clinicians not to harm their patients); beneficence (the duty to help one's patients); and justice (a duty to distribute goods in short supply fairly). The authors begin authoritatively: "This book offers a systematic analysis of *the* moral principles that should apply to biomedicine"[3] [emphasis added]. Their language invited the misinterpretation that persists today, that the authors were offering an exhaustive set of principles from which moral conclusions in bioethics follow deductively.

Yet, I would argue, this is not really what they were doing. The very paragraph that begins so authoritatively goes on to quote Tom Stoppard: "There would be no moral dilemmas if moral principles worked in straight lines and never crossed each other."[4] Principles in a true deductive system *would* work in straight lines; they could not contradict one another. A few sentences later the authors write, "Only by examining moral principles . . . *and how they conflict* can we bring *some* order and coherence to the discussion"[5] [emphasis added]. "Some order and coherence" is a much more modest goal than is complete and deductive certainty. In fact, it would have been impossible for anyone trained in philosophy after about 1950 to aim at the latter. Our training centered in epistemology, in the nature and—especially—the limits of human knowledge, in the failure of various quests for certainty across the centuries and throughout intellectual life. We were also trained, however, to pursue understanding in spite of uncertainty, trying to identify our assumptions, work carefully, and keep our conclusions cautious. Of course we sometimes fail in this. But our professional commitment is to careful, skeptical inquiry, as opposed, say, to the exuberant nihilism of some forms of postmodernism.

Beauchamp and Childress, however, were not speaking primarily to philosophers. Their audience included clinicians, who by the nature of their jobs need clear guidelines, and tended to interpret "principles" as rules. Their audience also included academics from many different fields, who had no reason to recognize the systematic doubt assumed as background. In addition, there are a number of statements in the book that invite being read as flat assertions. The result was a misinterpretation ("principlism") that soon became, and remains today, a target of attack. Bioethics has fallen prey to a common academic fault, in which certain criticisms and constructions remain conventional wisdom long after they have ceased to apply, and even if they were never really accurate.[6]

If practical ethics has never been simply the application of moral principle to concrete problems, then what is it? A more plausible position describes it as *moral reasoning* about practical questions: another kind of application, but considerably more open than the "application of principles" model. This interpretation, too, gives philosophy an important role. Philosophers can articulate the role that consistency plays in moral reflection, and are familiar with the sobering implications of some initially attractive principles. They should be skilled at clarifying concepts and setting out arguments. They bring the resources of our intellectual tradition to the table. (Thinking is often deeper when it takes into account what Aristotle, Kant, or Habermas have to say. Once we acknowledge that Shakespeare, Buddha, and Weber belong on this list, it becomes clear that all the humanities contribute to moral reasoning and reflection.)

This description—the application of moral reasoning to practical questions—is better. But it is not enough. It describes classroom teaching and academic writing and ignores the rest of what makes up the field. Even as a description of intellectual activity it is misleading, since it suggests that fields like law, epidemiology, and sociology are subordinate: research assistants who dig out relevant information rather than colleagues in framing and in answering the questions. The description also pushes doctors and nurses into something like a traditional patient role, people to be improved, possibly in spite of themselves. In short, the "applied moral reasoning" approach does not capture the concrete, collaborative, and interdisciplinary nature of the field.

Two Current Views and a Precursor

William Ruddick was one of the first to see things differently. In 1983, when "applied ethics" was still the dominant label, Ruddick identified "discursive moral competence" as a goal of ethics education: "the ability to discuss in appropriate moral terminology a variety of routine and rare cases with the variety of people likely to be involved in these cases . . . physicians, nurses, patients, relatives, lawyers."[7] His article fell by the wayside, but now seems prescient. Long before "discourse" came into intellectual fashion, Ruddick understood that one of our primary goals is to make it possible for people to talk with one another. Although this demands more than reasoning, and more than helping others improve their reasoning, it is not surprising that twenty years ago Ruddick focused on intellectual rather than social and psychological preconditions for productive conversation. Yet even there he was ahead of his time, understanding ethics talk more broadly than some still do. Ruddick identified "at least *three* modes of moral reasoning," which he simplified as "the Protestant . . . which stresses conflict of values . . . and agonizing decisions; . . . the Catholic mode, which supplies principles . . . [and] the Jewish mode, which uses anecdotes (actual or fictional cases) to pose questions and suggest answers."[8] (A twenty-first-century version of this list would probably include Islam and Buddhism. Ruddick's point was not theological, of course, but practical. Moral discussion takes a variety of forms, however these are named.)

Two recent discussions are still richer. Neither is purely theoretical; each turns to metaphor to capture the work of practical ethics. The first comes from Margaret Urban Walker, the second from Caroline Whitbeck. Walker's topic was not the field as a whole but ethics consultation. She described it as "a kind of interaction that invites and enables something to happen, something that renders authority more self-conscious and responsibility clearer. It is also about . . . maintaining a certain kind of reflective space (literal and figurative) within an institution, within its culture and its daily life."[9] Her point can be seen as a development of Ruddick's. He had allowed a variety of ways of processing ideas, of finding insight. For her, our goals include not just thinking ("ethics as moral philosophy") and talking ("ethics as discourse") but also awareness (of responsibility) and disposition (for reflection). And in linking metaphorical space to literal space, she moves bioethics outside the mind and into the world.

That last comment may seem unfair. Bioethics from the beginning has aimed at making a difference, in improving the way patients and research subjects are treated. But my second source, Caroline Whitbeck, points out that we often *describe* ethics as if it were a form of adjudication. In fact, she says, practical ethics is more like design.[10] Judges rule on the rightness or wrongness of what someone else has devised, where designers try to create solutions. Each situation presents a distinct set of demands, and the job of the designer is to create a solution that best meets those with which she is faced. Although there are clearly wrong answers, there is rarely a unique right answer. Anyone experienced in consultation, policy development, or exemplary practice knows that it is a form of problem solving, that ethical verdicts are never the whole point and sometimes simply beside the point.

Whitbeck's analogy could be further developed. For one thing, a designer must be responsive to parameters of many different kinds: the customer's desires, governmental regulations, the possibilities and limits of the physical world. An excellent designer will see possibilities and problems no one has thought to ask about. So in practical ethics we must be responsive to our client's desires, but also to the relevant law, and to morality. Our work goes beyond doing what is asked of us, a point to which I will return more than once. And designers do not just work on paper. They can create prototypes, look at problems in construction, and take a hand in designing the tools that will bring about the results they have imagined.

"Morality," like most terms, has a clear core and some disputed boundaries. Obviously it refers to the rightness and wrongness of actions, the virtues and vices of persons, the justice and legitimacy of political structures. I think of the moral attitude as fundamentally a matter of respect, but I do not use that term as Kant did. Respect, for persons and for the rest of the world, is a matter of recognizing, protecting, and nourishing whatever is of value. Under this description, morality cannot be adequately understood through philosophy alone. Equally essential are the other humanities and the sciences. Each in its own way helps us see what exists and understand its value. These, too, are points to which I will return in later chapters.

Practical ethics, then, is by nature interdisciplinary. The ability to see what is, to imagine what is not, to understand whether things can be changed—no single discipline has a lock on so much. The ethical thing to do has been defined as that which should be done, all things considered. This is a large and powerful phrase. During that week in November 1994, I both drew from philosophy and grew as a philosopher, but I also called upon the little I knew of anthropology, with its constant warning that descriptions are suspect. (Anthropologists sometimes call themselves "natural anti-scripturalists.") When I began each consult, I knew that the initial description I was given would bear only a rough resemblance to the fuller story. I had to know something about biology and medicine, as well. Some knowledge of law (which I had) and of mediation techniques (which I did not) would also have been relevant.

The Domain of Bioethics

Keeping moral space open, providing language and skills within it, identifying moral problems and helping create solutions for them: let us take these as a rough characterization of the goals that define bioethics. In the next chapter I will suggest a way of uniting these purposes still further: I will argue that our work may be understood as a mutual engagement, with our audiences and collaborators, in moral development. Together we learn to see, to think more clearly about, and to act so as to improve the moral dimensions within our domain.

For now, however, this loose preliminary description will suffice. The next task is to sketch the boundaries of our subject matter. After all, every field of practical ethics, whether in business, engineering, law, politics, or other fields, tries to maintain moral space, foster respectful conversation, and so on. But each has a distinct domain. For bioethics, this domain has largely been health care, health policy, and biological research.

The three arenas present an endless variety of moral questions from which bioethics has carved out a subset with an implicit central focus: the interactions between doctor and patient and researchers and their subjects. From the beginning I have been struck by the relatively narrow focus in bioethics, by the invisibility of whole categories of moral issues. Later I will call attention to the reasons for these boundaries and challenge their adequacy. At this point I want only to map them.

Bioethics tends to center around the doctor-patient interaction. I say "doctor" advisedly. Nurses do important work within bioethics, but much of their attention, too, focuses on what physicians should do. This discussion can empower nurses, who must occasionally, in conscience, resist doctors. But little in bioethics looks at nursing in its own right, as a separate field with its own problems. (In fact, most probably find that last sentence nonsensical. They cannot imagine nurses facing problems different in kind from those doctors face. "Competent patients have the right to refuse treatment; how could it matter which field is providing the treatment?")

Bioethics in the "dilemmatic" mode has focused on such questions as how much to tell patients, how strictly to keep their secrets, and especially, in every possible permutation, whether to allow or to hasten death. Literary scholars have examined the structure of medical records and case reports. Exemplary practices are usually explorations of ways to help dying patients. The methodological disputes of the past ten or fifteen years ("principlism" versus "narrative" versus casuistry, etc.) push the discussion one level back: rather than asking, "What should doctors do?," it asks, "What tools are most helpful in figuring out what they should do?"

One kind of inquiry moves away from dilemmas and principles of action to deal with the subjectivity of the patient and of the doctor (again, I use the word advisedly; there is astonishingly little on the inner lives of nurses). What is it like to be sick, to suffer pain and disability, to face death? Conversely, what is it like to open bodies, catch babies, to see the disease in a patient who feels well? Explorations of subjectivity aim at helping doctors better understand themselves and their patients.

Societal issues addressed by bioethics usually involve access and allocation: should the government ensure universal health care, and if so how, and how should we distribute our finite resources? Often access to health care is treated as access to a doctor, but certainly not always; and what is scarce may not be doctors but technology or organs. Within the past few years, attention has gone to "managed care": here the frustrations of doctors get attention, as do policies that limit patients' access to doctors, technology, and medication.

The ethics of biological research began somewhat differently, centered in the rights of subjects to be protected *from* (rather than to be included in) research; recently the discussion has broadened to include the latter.

In all these areas, discussions take place at different levels. In each area there is discussion of particular cases: Should Carlos's sister be told he has AIDS?[11] Were the placebo trials held in Africa justified? Should this child's liver transplant be funded? In each, there are also more general and theoretical discussions: What grounds the obligation of confidentiality, and how strong is it? What metaphors do people use as they describe being sick? What principles should we use in deciding who gets help? Discussion at any of these levels shapes discussion at others.

There are other subjects as well, flowing naturally from the field's historic core and changing its boundaries as they accumulate. New subjects arise from scientific advances (most recently in genetics), from the personal interests of influential people, and from repeated if dim glimpses of the fact that health care is not the only significant factor in health, nor doctors the only providers of care.

At the moment, then, the domain of bioethics centers upon health care, paradigmatically understood as a doctor helping a patient, but in principle including other health-care professionals as well. It includes the political and economic forces that shape medicine and access to it, and the biological research that underlies it. The domain shifts over time, growing in one area, shrinking in another, both acquiring and discarding foci. I will argue later that we need to take responsibility for this process, that we need to consider, and to reconfigure, the borders within which we operate.

Internal Goods, External Goods, and Virtue

In addition to the complex of goals and means that constitute a practice, each also makes possible unique forms of excellence. Chess players can conduct brilliant end games; architects can design buildings beautifully suited to their function. The equivalents within bioethics include a well-done ethics consultation; a soundbite for journalists that makes exactly the right points and does it memorably; an effective course design; a task force that does its job well; an article that is morally sound and practically useful; a good palliative-care policy; and so on.

In addition, there are what MacIntyre calls "external goods." His own example is that of a child who has been offered 50 cents for each game of chess she plays. Learning to play for the sake of the reward, she eventually takes pleasure in the game itself. Money is an external good; that is, it can be gotten without playing chess, and is not a necessary part of the game. Many other kinds of reward also count as external goods: they are not part of the activity, although their pursuit can promote excellence in it, as in the case of the child learning chess. More to the point, external goods can endanger the specific excellences of an activity. This double role is often conspicuous in professional sports: On the one hand, money and fame help motivate athletes to high levels of achievement. On the other hand, money and fame can lead athletes to destroy the goods that are meant to be part of sport. Competitors will sometimes cripple one another or themselves in order to win, or will accept a payoff and "throw" a game. External goods can destroy internal ones.

Bioethics offers some significant external goods. When Ron Carson and Chester Burns asked whether twenty-five years of bioethics have made the sick better off,[12] my instinctive response was, "I don't know. But *we* certainly are." The ways in which bioethicists benefit from doing bioethics deserve attention.

All professions offer some external goods, typically status, security, and a reasonable income. In most professions these rewards create at least some tension with internal goods. Junior faculty, for instance, can feel torn between the need to satisfy a promotion committee (an external good) and the internal goods of excellent teaching and research. For there can be a difference, unfortunately, between trying to please students and trying to teach them, even though the two may be closely linked.[13] Along the same lines, publishing for the sake of promotion is different from pursuing knowledge for the sake of the common good. Physicians also face conflicts: in a fee-for-service system, some will order tests or procedures not because a patient needs them but because they get paid more every time they order one. The motivation may be conscious or unconscious.

Bioethics offers higher salaries and larger stipends than do the standard academic liberal arts. As a newcomer to the field, I was struck by its relative affluence. During my first year in the field, I was invited to speak at another college; the arrangers asked what my stipend would be. I didn't know what to say, so I asked a colleague: "Would $35 be about right?" He was astonished. "Don't do that! You'll depress the market!" So, upon his advice, I asked for, and got, $350. For the "service" components of our work we often receive stipends five or ten times what a philosopher of comparable stature might ask. This comparison does not hold for medicine. If doctors are right in thinking that bioethics lowers their salaries, these particular external goods cannot threaten their integrity. The danger does exist for humanists.

In addition, bioethics can offer a certain degree of celebrity, if one is quoted in the media or does much public speaking. We can also have a certain amount of power, about specific cases or about policy, although the power is always diluted: our role is advisory, and usually we serve as members of a committee or a board. Finally, of course, there is the allure of security (in the form of tenure, when it is available) and of status within the profession.

So conflicts of interest of varying degrees are built into bioethics, as they are in all professions. The chance to gain money, security, prominence, or influence could distort what one chooses to say or do. This is obvious in dealing with the press, subtler in other situations. My colleagues and I have written elsewhere about interactions with the media.[14] Later parts of this book will explore some of the other conflicts. My major reason for identifying these external goods, however, is not to offer solutions for particular conflicts but to help in identifying virtues that may be of particular importance in bioethics, in part because of competing external goods.

MacIntyre's framework raises the possibility that individual practices demand distinctive virtues. (Something like "resisting countertransference," for instance, is probably a virtue necessary for psychotherapists, and not even nameable outside that practice.)

In exploring the role of virtue in the practice of bioethics, I will make no attempt to give a list. But I will argue that the lure of external goods highlights the importance of integrity and discernment, and that interdisciplinary work particularly requires intellectual virtues. I will describe an overlooked virtue that I can label no more simply than "choosing projects well." Finally, I will discuss how the communities we form contribute to or interfere with these virtues.

Purposive Practices

I asked almost all of those whom I interviewed whether they thought bioethics made any real difference in the world. "Yes!" said one, vehemently. Then, after a brief pause, she said more quietly, "It had better make a difference." There was another, longer pause. Finally, more cautiously, she ventured, "A lot of people don't think about things till it's too late. Patients, family, and staff." The implication was that her work helped them take up important questions earlier, when it might help. To my question another person said simply, "I have to believe that it does" make a difference. Almost all the responses I got fit within this spectrum. Often the answer was a kind of assertion of faith, and certainly a recognition that, unlike some of the arts, other humanities, and even sciences, bioethics is a failure unless to some extent it improves the world.

MacIntyre, writing about practices in general (certainly including the arts), takes a more modest position. He suggests that practices should fit together, each contributing in some way to the possibility of full, good human lives, if only for those who engage in them. David Miller, however, in response, makes a distinction: some practices have as an *internal good* the production of extrinsic goods: agriculture must produce food, medicine must, in some sense, heal.[15] This response provides another tool for the moral evaluation of bioethics: does it accomplish anything, and is what it accomplishes good?

These are large questions. Some of them I will address in this book, but others demand broad-scale empirical investigation and are beyond my reach. Some of the initial findings are not encouraging. A number of large recent studies, for instance, suggest that "advance directives" ("living wills" and assignments of powers of attorney for health care) make little difference in health care. These documents have been advocated for years as ways of changing some undesirable facts about dying in America: there is too much pain, certainly, and probably more intervention than patients want. On the other hand, the findings themselves, together with the debate about physician-assisted suicide, have at last led to serious attention to the way health care handles death, and the landscape may finally change for the better—not for the wrong reasons, exactly, but because of an unexpected sequence of causes. Questions about whether we make a difference are complex.

Although almost everyone to whom I talked believed bioethics makes some difference, most also recognized that we don't really know. The one clear exception was someone who did quantitative work. He answered without hesitation. "Oh, yeah, bioethics has made a difference. The FDA's 'Fast Track' decisions, for instance: bioethics gave a voice to the AIDS activists; they needed legitimacy from people outside the movement. And bioethics kept informed consent alive even when the courts were backtracking. Advanced directives, too. We have data showing how much the families felt relieved, knowing they had the advance directive to support their decisions."

Many who were less concrete, who said, in essence, "I have to think we make a difference," had teaching in mind. For many of us, teaching is fundamentally an act of hope. Year after year students pass through our classrooms and on into their lives. Sometimes, years later, we learn that those classroom hours mattered. Usually we never know.

Another sort of response was also common. "We contribute to a lively public debate. Sometimes we even create that debate. And that, in turn, shapes general awareness, policy, customs and individual choices." I believe this thinking may well be right, and that we are mistaken to evaluate the impact of bioethics solely by looking at some of its specific activities. The question is not just whether, say, ethics courses change the medical students who take them, but whether the now universal presence of ethics courses changes the fabric of assumptions in which medicine is practiced. Similarly, an ethics consultation service might not make a difference directly; the results in cases where consultation is sought might not differ statistically from the usual hospital pattern. But it is possible that patterns of practice in the hospital will change in the years the service exists, because of its sheer presence, the individual consultations it provides, and its educational activities. Obviously specific activities need to be assessed, and here again the preliminary findings are disquieting. Ethics consultations may reinforce medical power rather than challenge it, for instance.[16]

The same kind of question arises about our overall impact. If we are, for instance, creating and sustaining public debate, is the shape of that debate good? My doubts about the answer began during that week in November 1994, when I first suspected that bioethics helps divert attention from poverty, the most serious health problem in the United States. Europeans sometimes comment upon the American fascination with health care, in contrast to the considerably cooler attitude in Europe.[17] I have come to believe that bioethics both feeds upon and feeds that fascination. Henk ten Have believes Americans' interest will cool. Until that happens, and I see no signs that it will, several things concern me. One is the valorizing of health and "health behaviors," which sometimes seem our only secular values, and ones that are used to validate everything else. (I've seen friendship, sensuality, spirituality, and altruism all praised because they seem to contribute to good health.) Another concern I've already mentioned: the valorizing of medical care, and the consequent inattention to social contributions to disease. My qualms occur as a *leitmotif* throughout the book.

If bioethics were simply one intellectual discipline among others, its idiosyncratic shape might not matter. Each discipline has its own domain, and none can take on the whole world. But the practice called bioethics is more than an intellectual discipline. It helps attract and focus public attention; it tries to help society think more deeply and act more wisely about matters of health. For these reasons we need to ask whether we are making the world better, or dazzling it further into blindness. In contrast with the fora ordinarily available to scholars and clinicians, bioethics supplies a bully pulpit, one that we should use thoughtfully.

6 Bioethics and Moral Development

Questions about the proper goals of bioethics surface regularly on the electronic listserve maintained by the Medical College of Wisconsin. In one recent exchange, speakers took sharply opposed positions. David B. Resnick wrote, "I do not think the field itself should be committed to any particular ideology or political agenda. Bioethics is fundamentally a discursive, reflective activity carried out in the Socratic tradition. . . . Bioethics is philosophical reflection on medicine; it is not medical politics carried out under another name."[1] Stuart Sprague responded, "I find it interesting that the adjective 'Socratic' is used to describe a stance which is merely discursive and reflective, the implication being that it does not lead to any action or social change. We remember Socrates not because of the clever arguments he posed for his listeners but because of the impact those arguments had upon the social order, *including himself.* The friends assembled around him in David's painting at the Metropolitan probably wished that those discussions had not had so much impact. Perhaps we bioethicists should risk more"[2] [emphasis added]. I find myself in partial agreement with both speakers, but believe that the terms of the debate need to be recast. Bioethics is necessarily engaged in trying to change the world, but its tools are neither those of political activism nor those of pure scholarship. I will argue that we properly aim at moral development within ourselves and others, a development that comprises moral perception, reflection, and action.

My argument begins with the assumption that bioethics is a practice in the sense I've described, and that it has the loose cluster of goals to which I've pointed: keeping moral space open (Walker), providing language and skills within it (Ruddick), and identifying and helping design solutions for moral problems (Whitbeck). Most practitioners would agree with these goals. But, as the remarks quoted above demonstrate, many uncertainties and disagreements about our proper role remain. Those who serve on public commissions have been puzzled by their role: citizen, advocate, teacher? Each role legitimizes different actions.[3] The task force that developed national standards for ethics consultation had a hard time agreeing on the purpose of these consultations.[4] A literature and medicine course may have goals that aren't expressed on the syllabus, for instance, that students grow in self-knowledge. One might worry about the legitimacy of goals that are not disclosed.

Some of those with whom I talked described their activities as purely intellectual. The motivations for this stance varied. For one person, it was a matter of what interested him: "It's a good place to do philosophy. Just as in the 1920s physics was the best place to do philosophy. Medicine puts pressure on standard notions about persons, and so on, and that can spark creativity." For another, allegiance to the discipline in which he was trained demanded restraint: "I see myself doing descriptive work. . . . If I stood up and said, 'Don't do that egregious thing,' I'd feel I'd betrayed my discipline." Several people were most concerned with survival: "We don't take stands. We can't afford to be associated with one side of a controversial issue."

On the other hand, many voiced a desire to promote specific practical changes. Obviously that includes people involved in innovative practices like palliative care: "By making suggestions [for example, more morphine] to the nurses, we get a message to trickle up: the nurses suggest it to the intern. . . . Then the interns can use the info to look good in front of the residents, and thus up to the attending." But nonclinicians within bioethics also hope to bring about change. One ethicist, for instance, was working toward consensus statements about offering "futile care" within a large metropolitan area. Another, hearing a doctor describe a proposed genetic testing clinic and her fears about how insurance companies would use the clinic's findings, accompanied her on visits to the largest local insurers. Together the physician and the ethicist persuaded the insurers not to exclude anyone from coverage as a result of what a genetic test revealed.

Very often, however, those who wanted change hoped to accomplish it through others: "We have to remember that we're not number one; actual good will be done in the world through our students, whether they're clinicians or policy makers." "I followed a Peace Corps model: try to work yourself out of a job."

Not only does our clinical and policy work show this mixed and embodied set of goals. The same complexity typifies ethics education in medical school, and prebaccalaureate teaching as well. My undergraduate bioethics course explicitly includes the national ethical consensus, where it exists: students should know that patients have certain rights *even if* the students are unclear about the moral reasoning underlying that conclusion. I encourage these future doctors and nurses to treat patients in the ways now accepted as morally ideal at the same time as I help them learn to think productively about the many issues that remain controversial.

It should be clear by now that I think we are not simply scholars; our work is too practical for that single role. Both men quoted at the beginning of this chapter have part of the truth: We should not be ideological, as Resnick argues; on the other hand, we should acknowledge and seek an impact on the world, as Sprague contends. Can we manage both?

One way of understanding our work provides a middle road: we can see ourselves as engaging with our audiences in a mutual process of moral growth. That claim will strike some as trivial, some as repellent, and others as strange. These reactions, I believe, arise from common but inadequate understandings of what moral development is and what encourages it.

Those who find my suggestion true but unenlightening may think of moral development in the sense Lawrence Kohlberg meant it: as progress in reasoning about ethical dilemmas. It is obvious that we try to help our audiences, and one another, grow in this way. Those who find my model repellent may think of moral development as the kind of personal guidance and exhortation provided by parents and clergy; most of us explicitly disavow that kind of role. Readers who find my approach strange would say, perhaps, "I just don't think of my work that way." That kind of objection is significant, since those who do the work have an expert knowledge of what it is about; but that understanding can always develop. I believe the model of shared moral growth will provide a more complete understanding of our work than the "moral reasoning" model mentioned in the last chapter, and is more systematic than the metaphors of space, design, and discourse presented there.

My position depends upon the model of moral development that I present below, one that includes improved moral perception, reflection (which includes but is not limited to reasoning), and execution (that is, the carry-through from understanding to action). Most activities in bioethics can be understood as supporting growth of this kind, within individuals and in organizations, within ourselves and in those we serve. The process is a shared one, but that does not mean that each person gains the same thing in the same way from any given project. In some of my professional work I contribute analytic skills, and gain a knowledge of the life-worlds of those with whom I work; together we might construct a solution and press for its enactment. Another time I might contribute a concept from Aristotle, and gain a new way of thinking about the moral status of animals (from people who would never use that term). What I give and what I gain cannot be neatly separated into categories like principles and facts.

A brief history of developmental theory will be helpful as background for my own model. Although most readers probably think of "moral development" in terms of moral judgment (the Kohlbergian tradition), many different conceptual schemes concern the same basic questions. Religious traditions and psychological models, for instance, present notions of growth toward a maturity that could often be labeled moral.[5] For the most part I will stay with the stream of research that has been explicitly called moral development theory, which is considerably more complex than it often seems; but I will also touch on some areas not explicitly called moral development theory because they offer resources for thinking about my topic. Those fruitful fields include recent moral philosophy, cognitive development research, and concepts of "faith development."

The Piagetian Mainstream

The twentieth-century European and American tradition begins with Jean Piaget, who sought to identify universal invariant stages in a number of aspects of human development. His best-known work concerns cognitive development in children, but his work in moral development requires attention here. In that domain he posited two successive stages, the heteronomous and the autonomous. Heteronomy, a Kantian term, means being ruled by others, or by nonrational parts of the self. For Piaget, heteronomy consists in "unilateral respect for authorities . . . and for the rules they prescribe." Autonomy, which for Kant meant self-rule in accord with principles ascertained by reason, means for Piaget decision making based on "mutual regard among peers or equals and respect for the rules that guide their interaction."[6] (Note how different this notion is from the "autonomy" of the current discussion within bioethics.) Piaget's worry about excessive obedience, a concern widely shared after the Holocaust, has motivated a number of developmental psychologists.

In this country Lawrence Kohlberg, inspired by Piaget, articulated a fuller scheme differing in significant ways yet sharing important basic features, including a focus on moral reasoning. He also used the kind of instrument Piaget had used, which measures responses to briefly described moral dilemmas. Kohlberg posited three major stages of growth. In the first, "preconventional" stage, a young child thinks of right and wrong as behavior that will bring pleasing or displeasing consequences. In the second, "conventional" stage, a growing child or an adult thinks of right and wrong in terms of the expectations of her family, peer group, or country. In the final, "postconventional" stage, an adult tries to see the problem from a universal and impartial point of view.[7]

In her well-known challenge to this scheme, Carol Gilligan believed that she could identify a fundamentally different approach to moral dilemmas, an approach found more often among girls and women than among boys and men. Those speaking in this "different voice" sought to resolve dilemmas rather than solve them, valued maintaining relationships over a quasi-mathematical weighing of rights and duties, thought in terms of conflicts of responsibilities rather than conflicts of rights, and focused on the uniqueness of individuals rather than their abstract equality. Gilligan also found in this "different voice" an acknowledgment that disagreement is inevitable, and that complexity and uncertainty are inherent in moral life.[8]

About the same time James Rest, one of Kohlberg's students, expanded his basic scheme to include the *recognition* of moral dilemmas (which demands more than responding to a dilemma that is presented by others), as well as the decision to do the moral thing, and the strength of character actually to do what one believes one should.[9] In recent work Rest and his colleagues have made further revisions.[10] However, they retain and emphasize Rest's original insight that "morality is a multiplicity of processes."[11] That is, morality is more than a matter of thinking about moral dilemmas.

The intellectual line that began with Piaget, then—conceiving of moral development as moral reasoning, tested by hypothetical dilemmas, and progressing in clear stages through childhood and young adulthood—had by the late twentieth century broadened considerably. It sheds light on what we in bioethics are trying to do: to help clinicians and researchers increase their ability to recognize and think through hard choices, and carry out the decisions they reach; to help clinicians, students, and citizens respect disagreement when these discussions arise; to participate in efforts to resolve problems rather than simply solve them; to remind everyone, including ourselves, to think in terms of responsibilities as well as rights.

Recent Work in Moral Philosophy

Piaget and Kohlberg drew from the moral philosophy of their time, as have many other developmental theorists. Around 1980, moral philosophy began to open in significant ways, and it now provides material for a richer understanding of moral development. In mid-twentieth-century philosophy, ethics had been largely a matter of the analysis of moral language. In the 1970s, partly in response to the social tumult of the period, philosophical ethics became newly practical. It addressed issues like racism, civil disobedience, and nuclear war. At first this discussion unfolded as a contest between utilitarian and deontological perspectives, each of which claimed to provide the one best way to answer questions about the right thing to do.

By the early 1980s this rather tight conceptual program had opened, not to say exploded. Alasdair MacIntyre made virtue a central topic; Lawrence Blum argued for particularity as, at times, a moral accomplishment rather than a defect; Bernard Williams pointed to the anomaly of "moral luck" and to the importance of whole lives, not just particular choices.[12] Moral philosophers began to realize that the field's return to real questions about real lives had been narrow; there was a lot more to consider than whether particular actions were justified. Edmund L. Pincoffs put it this way: "[T]he problem of moral education is not so much teaching [people] how to make moral decisions as giving them the background out of which arise the demands that decisions be made."[13] In fact, he argued, "the primary aim of moral education is to encourage the development of the right sort of person."[14] Pincoffs rejected reductivist descriptions of the right sort of person. He or she cannot be simply described as acting always out of respect for moral law, nor as one dedicated to producing the greatest happiness for the greatest number of people. "Once we recognize that ethics is not best reduced to the search for a universal solvent of moral dilemmas, we are free to explore the varied and rich texture of moral talk and thought."[15]

These insights are important for concepts of moral development. Of particular use is the work of Lawrence Blum, who draws from psychology and education as much as from philosophy.[16] Blum's concept of "responsiveness" makes use of a number of findings about human psychology: how empathy is acquired, differences in the ability to understand oneself and others, the sense of relatedness and its lack, and the activity of making sense of a situation. Blum argues that fully understanding others and one's relationship to them is not just a precondition for moral action, but a moral enterprise in itself. It demands, after all, compassion, attention, perhaps courage, and other virtues as well.

Such newly complex understandings of the moral life contribute to a more satisfactory account of moral development. Later I will draw from them to support my own account and my claim that practical bioethics is a participation in such development. For the moment I can at least note that responsiveness (as Blum calls it) toward patients and colleagues is learned, and that the work we do as bioethicists can promote or impede it.

Cognitive Development Theory

There are other resources from which to draw, less explicitly concerned with moral growth. One is *cognitive* development theory, as it has evolved since the time of Piaget. He, studying children's understanding of the physical world, formulated a progression from "concrete" to "formal" thought, the latter an ability to deal with abstractions and logic.[17] Some of his intellectual descendants argue for a further stage, a "postformal" one, necessary in understanding other people rather than the physical world. This stage is described in terms of complexity and openness to contradiction.

This complexity has been formulated in superficial—and dangerous—ways, but also with sophistication; in that form it is useful. Cruder formulations peg cognitive immaturity to measures like failure to consider all perspectives "valid in their own right" and to "strong reliance on factual evidence."[18] Taken literally, these criteria are alarming: they seem to obviate rational discourse, which demands getting the facts as right as one can, and rejecting some perspectives as mistaken. But it is important to see even these unsophisticated formulations in context. The writer whom I just quoted worries that the attitudes she calls immature (reliance on evidence, belief that some perspectives are invalid) make young thinkers "more vulnerable to external agencies that define what is objective truth: authority figures, institutions, society."[19] She also writes that such thinkers may not be able to understand others as having worldviews and being worthy of respect, and might greet disagreements with simple assertions of right and wrong.

Other thinkers present postformal thought more carefully. One calls the goal "dialectical thinking" and contrasts it to objectionable forms of relativism: "In a contextual/relativistic world view, random change is basic. . . . Prediction is impossible, as all people and events are unique and continually change in unsystematic ways. Consequently, contradiction runs rampant. There is no order." In contrast, a dialectical (or organic) worldview simply recognizes that things are always changing, solutions are not eternal, and there are always new perspectives and new facts with which to deal.[20] This construction of postformal thought fits nicely with Gilligan's perspective, and highlights another aspect of what we are trying to accomplish in bioethics: that we and our audiences deepen our recognition of others as interpretive thinkers, makers of meaning; and that we keep clearly in mind our own limitations as knowers.

Faith Development

Finally, I would like to touch on one other field from which we might glean insights into the idea of moral growth: what is called "faith development" theory. Its concepts can be useful to even the most secular of bioethicists, once we understand the broad but humanly deep sense in which it uses the word "faith." The discussion takes place largely within liberal Protestantism; James W. Fowler's *Stages of Faith* began the field and remains its center. He defines faith broadly, in terms of the meaning we make of our world, especially after we abandon the certainties of childhood. Faith refers to holding some things to be fundamentally (perhaps transcendentally) of value, not just to finding them true. ("Credo," in fact, usually translated as "I believe," is etymologically tied to such phrases as "I set my heart upon.")[21] Under this interpretation, "faith" resembles Blum's "responsiveness," an attribute that combines awareness, concern, and a tendency to act.

As its name suggests, "faith development theory" grows from the Piagetian-Kohlbergian stream. It deserves our attention for a number of reasons.[22] For one thing, it addresses adult development, whereas much of the work in cognitive and moral development does not. For another, Fowler enjoys certain rhetorical advantages because he writes within a religious tradition, and these are advantages to anyone seriously interested in moral issues as such (rather than as intellectual puzzles). Fowler's religious tradition provides a sense that one must attend to the important things, whether or not these can be said precisely or measured at all, and values the use of metaphor and myth. These devices can be powerful spurs to thought, a point appreciated by William Ruddick's concept of "discursive competence." In terms of a distinction drawn from philosophy of science, theological work that is intended to provide *justification* to those within a certain tradition can provide for those of us outside it a context of *discovery*—discovery of significant elements of human experience.

Fowler, for instance, points to what he calls the "sacrament of defeat." He finds it crucial as we move toward maturity: "unusual before mid-life, [this stage of development] knows the sacrament of defeat and the reality of irrevocable commitments and acts. What the previous stage struggled to clarify, in terms of the boundaries of self and outlook, this stage now makes porous and permeable. Alive to paradox and [to] the truth in apparent contradictions, this stage strives to unify opposites in mind and experience. It generates and maintains vulnerability to the strange truths of those who are 'other.'"[23] This is a more resonant description of what the cognitive development theorist calls postformal thought. The word "defeat" suggests the emotional magnitude of what must be accomplished here, and "sacrament" suggests that it is a moral accomplishment. Several points for bioethics arise from these reflections. The most basic is that we and our audiences are not just resolving particular questions but modifying and reconstructing worldviews. Another is that recognition of mistakes and vulnerabilities is an important part of doing so.

In sum, various theories of adult development present us with different, but complementary and even convergent, perspectives on the tasks of adulthood. They demonstrate that moral maturity is a more complex achievement than "thinking for oneself" or "taking an impartial stance." Making use of these resources, I can sketch a notion of moral maturity and the paths that bring us to it. It is only a sketch, meant to give substance to and make plausible my claim that bioethics is an engagement in the process of moral development.

Again, I include within bioethics the other medical humanities; the considerations I've just given should underline quite powerfully the need to do so.

A Framework for Thinking about Moral Development

The core of moral development, I suggest, is the ability to recognize and value properly persons (including oneself) and the social matrix that partly constitutes human life. In principle I also include here a moral appreciation of animals and of the environment, and both are relevant to bioethics; but I will focus on the first two. I will be speaking of something like Blum's "responsiveness," an attitude that comprises knowledge, appreciation, and a disposition to act, but I will sometimes use terms like "recognition," "appreciation," and "respect" interchangeably with it. This stance requires moral reasoning, since the world presents us with a complex tangle of good and bad, actual as well as possible. Responsiveness also implies action, obviously in particular situations but also, I will argue, as a matter of habit; that is, I want to blend Blum's responsiveness with Aristotle's virtue: a habitual disposition to act rightly, out of an understanding of what in the world around us is worth fostering and preserving.

Responsiveness to Persons

Appreciation is most basically due to human beings, to their capacities to flourish and to suffer. The more sensitive one is to the effects of one's behavior on others, the more one is likely to help.[24] Similarly, understanding that people can make responsible choices is bound up with allowing them to do so and offering help in terms of information, time, and emotional support.

Becoming a sensitive observer is a complex task. Empathy (defined as distress at the distress of others) can be identified in very young infants: newborns cry when they hear other babies cry. But this infant response needs to develop. Rest notes that there are "striking differences in . . . sensitivity to the needs and welfare of others. . . . [T]he capacity to make inferences about the needs and wishes of others develops with age."[25] Not age alone, however, but age together with experience and education. One must learn to "identify the . . . meaning in the behavior of several [people] who are interacting with each other; infer what their respective wants and needs are; and imagine what actions might help."[26] (This is the sort of activity which Blum casts as a moral achievement.) Development toward this stage is gradual. In the presence of others' distress, infants simply feel distressed themselves. A young child will take a further step and try to help, but what she does is shaped by her own needs rather than those of the other person. (She might offer a crying toddler her own doll.) Later one comes to understand that other people have particular, individual needs.[27] Those familiar with health care will recognize the possibility of regression instead of advancement; the possibility that learning anatomy, physiology, and diagnosis make it harder for doctors to see the patient as a person; the tendency to offer patients what the doctor would want rather than what is appropriate in light of the patient's values.[28]

A less-remarked-upon aspect of this facet of moral growth is *self*-awareness and understanding. Although it might seem that we need no help in understanding our own purposes and our own suffering, thinkers as different as Freud and Wittgenstein argue otherwise. Within health care, David Hilfiker has written poignantly about the cost to doctors (and ultimately to their patients) if they have no chance to deal with their own anger and grief. The cost is not simply emotional; it is moral as well: as I will argue in chapter 10, virtues like humility depend upon a kind of self-knowledge that is not simply intellectual.[29] Unless we know our own emotions, we are likely to be driven blindly by them. Compassionate self-awareness is part of moral growth, and not just because it helps us treat others better. We ourselves are worthy of respect.[30]

Moral growth also involves expanding our moral community, our understanding of whose distress, or whose joy, counts. (This expansion, incidentally, is at the core of the progression in Kohlberg's stages.) It is natural for even young children to care about their family and friends. As children grow, ideally, that concern extends to outsiders; finally, with luck and education, we recognize that all human beings, even those most distant and foreign, have lives that matter. Again, research bears out the connection between our understanding and our actions: The broader one's moral community—that is, the more people (and animals) one understands as having lives that matter—the more one is likely to help. Samuel and Pearl Oliner studied hundreds of Christians who had rescued Jews during the Holocaust, and found that the rescuers had learned from their parents to be "significantly more inclusive" in their understanding of to whom they had moral obligations. In contrast, nonrescuers' parents had inculcated more limited obligations, to family, friends, elders, church, and country, but not beyond.[31] This also shows within health care: The race and class lines that divide American society are reflected both in the quality of care received by people of color and the poor, and in bioethics' relative neglect of these issues.

Appreciation of Social Structure

If moral development demands an appreciation of others as persons—as complex beings with inner lives, capacities, and goals—it also demands an appreciation of the way these lives interweave, of the complex interdependence that makes up the social world. Human beings are social animals; the language and customs that give meaning to the world are inherited, transformed, and bequeathed. Even a hermit has first learned to be a person by living with others. Most of the good things in life are possible only because people work together, share the burdens, and play by the rules. So an important part of the moral life is coming to see, value, and contribute to cooperative institutions: "to understand the nature and function of social arrangements (e.g., promises, bargains, role-defined divisions of labor . . .)."[32] Just as the inner lives of persons can be malformed, so can institutions. In both cases a moral life demands that we *see critically,* in a way that makes us ready to act in accord with what we understand. ("Ready" but not automatically committed, since the moral world makes a variety of conflicting claims upon us, including a reminder that not every problem is everyone's business.)

Again, experience and education are crucial in supporting this growth. Fortunate people know from experience that cooperation and mutual respect are possible, that communities work, and that life within them can be good. Lawrence Kohlberg tried at one time to develop abstract moral reasoning with prisoners and delinquents, but found that their experience of actual human societies made them skeptical of the possibility of cooperation. As a result, he tried to establish "just communities" within the prison itself. His hypothesis was that "people who are cynical, self-protective and brutalized need . . . to experience that their contributions to the community are reciprocated . . . [and] that cooperation is a workable—and even preferable—way to live."[33] Medical training, arguably, needs similar corrective measures. I don't want to exaggerate this analogy, which could easily be insulting; there is a world of difference between most doctors and most felons. But many people believe that medical training produces doctors who are, like the prisoners Kohlberg described, "cynical, self-protective and brutalized."[34] Furthermore, many observers have noted that doctors finish training ready to practice independently, but without a sense of shared responsibility: that is, without the habit of seeing the social matrix of which medicine is a part, without a commitment to evaluating and trying to shape it.

Some examples: the idea of allocating resources in a socially responsible way was slow to gain ground; doctors felt deeply obligated to their own patients and not at all to others. Doctors are notoriously slow to take action about impaired colleagues, and when they do the action is likely to be individual—refusing to refer their own patients to the physician, rather than taking steps to protect all patients from him.[35] Doctors can be oblivious to what nurses and others on the "team" actually do.[36] The AMA and its state analogues are often thought to seek the welfare of doctors rather than the good of the public.

Nurses, for their part, can be blind to social fabric in different ways. Although they are trained to be sensitive to the family and home situations of individual patients, this is only part of what needs to be seen. Nurses often define their duties narrowly, and find professional organizations, lobbying, and so on, irrelevant—yet (for instance) if good hospital care demands adequate nurse staffing ratios, *only* collective action will preserve them. (Competitive pressures, and the fact that nursing salaries are a large part of any hospital budget, will force these ratios down unless something like regulation levels the playing field.) And, like doctors, nurses can be blind to the ways in which hospital culture makes patient privacy a joke, and respect for patients sometimes a facade. Most moral agonizing by health-care professionals concerns choices about individual patients rather than the moral character of custom and institutional structure. Bioethics has of course been shaped by and contributes to this focus.

Moral Reasoning and Reflection

The more fully one appreciates the social structure, the more thought becomes necessary. We face conflicting perspectives and competing obligations, and must often choose between, for instance, hurting someone and being fair. So moral maturity involves thinking through moral complexities. Pincoffs points out that attention to a certain kind of complexity—to dilemmas, or as he calls them, quandaries—*results* from moral sensitivity, rather being in itself the soul of moral life.[37]

Within practical ethics, the academic field that began around 1970, the term "moral reasoning" is conspicuous. It is used to emphasize that we can and must think about moral questions, that our values are not simply matters of socialization and taste. I like to believe that this particular point has been made, at least within bioethics; people expect to find reasoned positions that will help them think through their own. Perhaps I am too optimistic. But in any case I think it is time to broaden the term to "moral reflection." "Reasoning" suggests formal, explicit thought; "reflection" is broader and more open, allowing for the variety of ways in which we come to see and to understand. The emotional content of an experience, for instance, can reveal moral content.[38] Human beings have a richer emotional life than animals *because we have higher cognitive faculties*.[39] For these reasons "moral reflection" is a more accurate description of a central dimension of moral growth.

That is not to call "moral reasoning" narrow. Reasoning is never a simple matter of deducing conclusions from premises. In ethics, reasoning includes formulating principles and testing them for coherence; considering their implications; wondering about their provenance; and searching for deeper or more general considerations. It includes the use of analogies and questions about reversibility ("How would I like it if . . .") and about generalizability ("What if everyone . . ."). It takes into account long-term consequences. It is a matter of developing a richer and more precise moral vocabulary, so that, for instance, "wrong" can be made more specific: Are we considering the claim that something is absolutely wrong (never justified) or prima facie wrong (justified if it is the only way to avoid more serious wrong)?

But moral reflection is still more expansive than moral reasoning. Among other things, it includes the consideration of myth and metaphor, respect for one's own nagging doubts and suspect enthusiasms, and the use of moral imagination. This last consideration makes another person's point of view more vivid, and helps us think of creative alternatives to what seem like dilemmas. It makes us think about the conditions that led to the quandary, to do what is called "preventive ethics."

Finally, taking a clue from the work in cognitive development summarized earlier, I would identify cognitive complexity as a part of moral maturity and an intimate part of moral reflection. Cognitive complexity is a disposition to see the world as constantly in process; to appreciate that each human being interprets the world and that all knowledge is mediated (through sense perception, language, social location, and so on); and to understand that one's own knowledge is shaped by one's own interests and situation. It is a matter of not only welcoming but seeking out new perspectives. But it is not the nihilism that makes all points of view equally valid.

As desirable as these skills in reasoning and reflection, and this ability to see the contingency of knowledge, might seem, it may not be clear that they themselves count as *moral* excellences. I count them as such for a number of reasons. First, they are a precondition for most moral action, by making an adequate understanding of situations more likely, especially of the deeper needs of others. Even so, these skills may seem to be no more morally relevant than good eyesight or sharp hearing, instrumentally important but not intrinsically moral. Indeed, there are many within philosophy who claim that personal goodness and the ability to do good moral philosophy have nothing to do with one another. There are several reasons, however, to consider these abilities more than instrumentally good. To begin with, they are acquired excellences, taught and learned, and only possible for persons. More important, some of the traits that help develop them (and partially constitute them) are themselves virtues: humility, for instance, and courage (the ability to face the possibility of being wrong, or being indefinitely in doubt). In addition, these abilities, particularly moral reflection and cognitive complexity, contribute essentially to respect for others and for oneself. Physical perception, in contrast, complex and conceptually laden though it may be, does not tell us that other people are makers of meaning. Finally, these intellectual habits contribute to the ability to take responsibility for one's own actions, because they include the disposition to reflect on the nature of knowledge and the consequent ability to resist claims to authority. Skills in moral reasoning, the habit of moral reflection, the attainment of cognitive complexity—all these are part of what we mean by moral maturity.

Action and Virtue

So far, this framework maps roughly onto the first two of James Rest's "four-step" schema: perception, reasoning, commitment, and carry-through. Some readers may believe that the last two are inappropriate aims for bioethicists; in fact, that is where the crucial part of my argument needs to be made. But for the moment I need to put that concern aside. For now, I want to point out that although my divergence from the Rest schema becomes greater after his first two steps, the difference is a broadening rather than a disagreement. I will be talking not only about specific actions, but about the habitual disposition to act, which is a virtue.

Rest's analysis is basically action centered: he is asking about the psychological components of eventually doing the right thing. (To many the word "bioethics" has the same kind of connotation. "Ethics" suggests, in ordinary language, a set of right actions, and of course bioethics has often centered on difficult, discrete decisions: when to pull the plug, how to allocate scarce organs, and so on. But many use "bioethics" more broadly, and I want to encourage that usage.) Virtue ethics takes a different question as central: What does it mean to be the right sort of person? Ideally that question is embedded in a larger one about the meaning of a decent society.[40] The central task of moral development, then, becomes not attaining the skills to see and solve hard ethical problems, but growing toward the kind of person who understands and properly values, in life as well as in thought, living beings, especially human beings and their institutions. Such a person will still face hard questions, but the ability to recognize and reason about them will be only part of a full moral life. It is obvious that skills in moral reasoning, for instance, may be necessary for moral maturity but are not sufficient for it.

Rest uses the term "ego strength" to explain what is missing. In less clinical language ego strength might be called character, the ability to do what one believes is right. Neither ego strength nor character is a temporary thing; both fit naturally with the language of virtue. A virtue is "a deep and enduring acquired excellence of a person, involving a characteristic motivation to produce a certain desired end and reliable success in bringing about that end."[41] The "end" might be coping with danger, finding the truth, helping those in need, and so on. Virtues include emotions, for the virtuous person does what is right willingly, not with gritted teeth. They are acquired gradually. Taken together they are an essential part of *eudaimonia,* Aristotle's term for a full human life. Since for Aristotle this life was lived in a community—in a household, and in a civic society—virtues as a set constitute "a kind of social ability whose value lies in their promotion of the end of living in a community in which all have the opportunity to live well."[42]

"Ego strength" suggests courage and determination, rooted in psychological states like personality integration and a sense of identity. Perhaps we could say that, just as a prism diffracts light into visible bands, virtue language pulls from "ego strength" a number of separately attractive components: courage, generosity, persistence, patience, and so on.

Moral maturity, then, as sketched here, involves perception, emotion, habitual action, skills in reflection, virtues like courage, patience, and perseverance, and more. These qualities are acquired gradually. The communities in which we live help, and hinder, our development of them. The work of a field like bioethics can contribute to their acquisition.

Moral Development as a Goal for Bioethics

I can now turn to the central question of whether the encouragement of moral development is an appropriate goal for bioethics. Resistance to the idea will not concern the first two elements identified above, moral perception and reflection. Helping one another and our various audiences see and understand is clearly what we are about. But many will believe it is not our business to go beyond that, to help people actually do the right thing, and do it willingly, habitually, and with understanding. The doubts spring from concerns about misuse of classroom authority, about the special role of scholarship, and about the limits of our own abilities.

Yet it is obvious that we often aim to bring about change. The very choice of one project over another, whether it is writing a paper or forming a task force, is often motivated by the desire to see something improve, within a student, a hospital, a professional organization, a legislature, the country. Furthermore, we contribute more to those projects than moral analysis. One of the people with whom I spoke reported, "I went to M&M [Morbidity and Mortality Rounds, where errors are reported and analyzed] regularly for years. I always asked, 'Was the patient told about the mistake?' Eventually it became routine in the department to tell the patient or family. I didn't have to ask any more."

I heard many other examples of bioethicists contributing to moral growth, broadly construed, in those they served. One ethics center put together a task force to develop a proposal for the legislature. The subject matter was "drive-by mandates" (state laws forcing managed-care organizations to pay for certain things, like experimental cancer therapy or two days in the hospital after childbirth). The center created the task force because it believed the current issue-by-issue legislative response was both inefficient and unfair. The ethicists brought together influential interested parties, whose conclusions were likely to be respected. This center hoped things would change, and in a certain direction, but it worked for a process rather than a single "cause." Part of what the members hoped for was change within the task force participants, perhaps by broadening their respect for "competing" groups, and certainly by sharpening their skills in productive moral conversation. If the project succeeded, legislators, too, would be able to see and to act upon a wider and fairer agenda. All this counts as moral growth, in perception, reflection, and ability to act.

I see this kind of activity in CEHLS regularly. When Len Fleck organizes public dialogues on issues of justice in health care, he does not only describe and recommend what he calls "rational democratic deliberation," he also helps it happen. He creates scenarios and leading questions, then guides the group in exploring the issues and listening to one another. At the end of the session, participants understand the issues more deeply, have a clearer notion of what democratic deliberation is and why it has value, and have gained skills in participating in it. When Tom Tomlinson and Diane Czlonka wrote about how to develop a hospital "futility" policy, they were not simply analyzing concepts and values. They were suggesting concrete steps toward a practical solution, against the background knowledge of how hospitals work and clinicians think.[43] When Howard Brody chaired a state-wide commission on physician-assisted suicide, he was not only helping people think clearly, he hoped to help bring about democratically respectful and morally adequate legislation.

In working with the media, my colleagues and I have argued elsewhere, we should sometimes deliberately speak for those whose voices are unlikely to be heard—so that valuable perspectives can be heard, but also to encourage the habit, in ourselves and in our audiences, of listening for those neglected voices.[44] This practice, too, counts as moral growth.

The kinds of encounters that are lumped together under the label of "ethics consultation" and sometimes caricatured as "giving advice" actually aim for and accomplish a variety of ends. One of the people with whom I talked worried on just these grounds about the effort to credential ethics consultants; she believed that doing so would be reductive, eliminating the variety of activities it now involves. She mentioned in particular the therapeutic and educational aspects of consults. She was using "therapy" in the informal sense; most of us have had the experience of simply listening to someone agonize, not being asked even for clarification, let alone advice—and without anything helpful to say if we had been asked—but knowing that the listening had helped. We are not and should not try to be therapists, but this basic, human, healing interaction is appropriate.

Most of my interviewees implicitly agreed that ethics consultations have a variety of means and ends. To begin with, they almost always tried to improve the moral quality of the conversation, and doing so usually involved process as well as content: "I bring in other voices, provide context, information, talk about how people have discussed the issue other places. Facilitate discussion. I don't give advice, don't say, 'This should be done.'" Sometimes what ethics consultants hope for is increased skill in moral reasoning: "My greatest joy is in hearing surgeons in M&M carry on the moral conversation themselves, in sophisticated terms. I don't even have to bring the ethical questions up." But often the goals were still wider: "What I do at the hospital makes a difference. Especially conferences about cases that are distressing people. *They learn to respect one another's points of view;* they learn some of the relevant law; *they learn to listen to one another.* I tell the medical students: You'll learn a spectrum of positions, and be able to stake out a position for yourself, learn that other people are other places; and [you'll] *know that on some of it you'll change your position later. You've got to be humble*" [emphasis added].

This last point brings to mind Fowler's "sacrament of defeat," as well as the self-understanding that is an essential part of moral perception. A similar goal is implicit in the words of someone teaching in a medical school who held "ethics conversations" off-campus once a month. During them he would present a case, which was always followed by "talk and talk and talk." He described the environment as safe and "family-like." "I absolutely believe in the power of reflective conversation. . . . When you have safety you have more power to reflect, and to be vulnerable." This man was clearly concerned with more than moral logic.

Similarly, those who bring medicine and literature together want to promote reflection on the part of clinicians and society about the meaning and experience of health care. Professionals in the field of medicine and literature will usually disown the label of bioethics, but in the sense in which I use it, they belong.

There are many other examples. To encourage the ethical practice of health care, we have students rehearse certain skills, from helping patients make decisions to confronting a fellow student who cheated. Having these skills makes the student more likely to act in a morally desirable way. Occasionally we exhort people to action, as I have done in public speaking: "Every church group in the country should petition Congress to have their health-care benefits taxed."[45] I do this to make a point about injustice, and of course I present what I take to be a persuasive moral argument to that effect, but my intention is educational rather than activist. Nevertheless, I would not mind being taken up on the exhortation, and once I was.

We also implicitly aim at moral development when we encourage and practice self-disclosure. One of my respondents tells her medical students of a time when she cheated, emphasizing that decades later she still remembers and regrets it, hoping that her story will help them develop character. In another chapter I will describe other ways in which revealing one's own errors can help others grow.

Some of this may sound paternalistic and patronizing, what someone has called a "deficit" model, criticizing it in bioethics as well as in medicine: "'They' won't recover, or grow morally, or be empathic unless 'we' do something to them." But what has become clear to me is that even in the classroom, where this attitude is most likely to be appropriate, the teacher is only making things available—knowledge, understanding, and skills which students will accept or reject. Furthermore, she is learning along with the students, and some of what she learns amounts to moral growth: the habit of listening; of attending to students, their problems, projects, and worldview; of discovering morally relevant aspects of cases and issues. Without this constant growth in respect for others, our work fails, as one of my interviewees found: "There was a big lawsuit in obstetrics. I went to the chair of the department and said, 'If you'd talk with your patients more, none of this would happen.' That went over like a lead balloon. They held a department meeting and invited me, listened to me; then the chair said, 'Anyone agree?' No one did. I've been *persona non grata* there ever since."

This aspect of engagement in moral growth, the fact that the ethicist, too, must be constantly growing, is something I will explore further in a later chapter. At this point let me simply offer this analogy: clinicians repeatedly talk about learning from even the most stricken of their patients, from the "discarded" demented elderly, from the dying, from the seventeen-year-old single mother. "It's a miracle [young women like this] are even alive," said one of my respondents. "And they respond so eagerly to any encouragement. If we could just harness their ability to endure." She learned something about courage and persistence from these patients. Similarly, David Hilfiker once spoke of discovering his own mental illness (depression) and asking simply to live among the poor at the home he had once helped administer, to contribute labor but not medical or administrative help. He asked the other residents not to inquire further, and was struck by how simply accepting they were: "We noticed things haven't been right with you, Doc." There were no questions, no prying, no special attention. The men assumed, in Hilfiker's terms, that there is brokenness within each person and each life. He came to recognize that the experience of failure, and the acceptance of it, is part of becoming a full self.[46]

Something similar should be true of work in bioethics, but with this important difference: the clinicians I've just quoted were learning in spite of being greatly advantaged in comparison with their patients. In bioethics, in contrast, where we do not so much work with the sick as with those who do (and those who draft legislation, conduct research, etc.), we rarely have this kind of advantage over those we serve. That makes the opportunity for mutual growth still more obvious.

My claim that a defining goal of bioethics is an engagement in moral growth (within certain domains) is probably surprising. But I believe this goal, richer than that of "moral reasoning" and "informing the conversation," captures more fully what we do. It also helps resolve the dichotomy with which this chapter began. Engagement in mutual moral growth is more than simply academic rumination, but it is not the same thing as activism. Activism aims at a specific end and uses any means that are likely to work. Morally acceptable activism will exclude deception, coercion, and so forth, but will not shy from effective sloganeering and horse trading. The purpose of activism is to bring about change, and increased public understanding and virtue are sought only as means, if at all. Mutual engagement in moral growth must meet considerably higher standards. It aims at strengthening perception and understanding, never just at bringing about a certain action. There is always, consciously and deliberately, space for reconsidering ends. Within those limits it does not exclude political commitment and action, but no specific commitment is ever its defining purpose.

Thinking of our work in these terms may help us work more systematically and more imaginatively toward ends we already implicitly endorse.

7 Virtue in Bioethics
Choosing Projects Well

Suppose, then, that bioethics is something like a practice. Its goals may be thought of as a loose bundle (keeping moral space open, designing solutions, promoting a better public discourse) or in terms of the model I've just offered, as a mutual engagement in moral growth.

However we describe our goals, certain virtues are necessary to accomplish them. When I asked people what virtues they thought were important for the work, the most common answers referred to open-mindedness: "flexibility," "the capacity to have multiple perspectives, to listen carefully." My question about what vices, if any, particularly impair good work was often answered correlatively: "rigidity," "dogma." Many of the people with whom I spoke were thinking in terms of individual encounters, but Larry Churchill has made the same point more globally. He argued that we must maintain "skepticism about our usefulness, agnosticism about our answers, pragmatism about our usefulness, empiricism in our endeavor, and irony in our use of 'expert'"—together with utter seriousness about the task.[1]

One could call each of those attributes a virtue. In naming virtues, one need not stick to a classical vocabulary: one might instead include something like adventurousness, for instance. One researcher found those who rescued Jews during the Holocaust to be generally adventurous.[2] Nor need we draw only from the Western tradition: Buddhism has interesting things to say about compassion, and the Confucian scholar Mencius thought about emotions in quite a different way than Aristotle did.[3] (Emotions are partly constitutive of virtue, a virtue being a disposition to act in the right way for the right reasons.) Even within our own tradition we need not strain to find a conceptual rigor that is not there. Zagzebski points out that "our language does not contain a sufficient number of names that convey the full unified reality of each virtue. Some names pick out reactive feelings (empathy), some pick out desires (curiosity), some pick out motivations to act (benevolence), whereas others pick out patterns of acting that appear to be independent of feeling and motive (fairness). For this reason it is easy to confuse a virtue with a feeling in some cases (empathy, compassion), and with a skill in others (fairness). The result is that it is very difficult to give a unitary account of virtues using common virtue language."[4]

Taking advantage of the freedom this description offers, I will speak rather loosely of some virtues that I believe are required in bioethics. Some of my points could be made under different rubrics; the topic of this chapter, about taking on the right projects, might be approached as a matter of role obligation.[5] I stay with the virtue model because it naturally leads us to think in terms of whole persons and whole lives, rather than specific actions. And of course a focus on virtue follows naturally from considering bioethics as a practice; MacIntyre's major purpose in introducing the concept was to provide a contemporary way of thinking about virtue.

Many different virtues are required of the bioethicist; there is no algorithm for producing a list. Many of us teach, an activity that demands humility, courage, and a sort of cheerful selflessness. In all of our activities, we need to listen, and Joseph Beatty has argued that the ability to do so well is a virtue of major significance.[6] The people with whom I spoke offered openness, compassion, fairness, tact, truthfulness, and "a certain idealism, a willingness to deal with the issues society doesn't want to: either go deeper, or actually take a stand." For his part, MacIntyre would argue that honesty, courage, and justice are necessary for bioethics as they are for all practices, and I do not disagree.

My own discussion will focus on choosing the right projects (which I consider a virtue in itself), and the way internal and external goods place pressure on the effort to do so; on the moral demands of interdisciplinary work; and on what it means to be a virtuous community, one in which the goals of the practice are likely to be accomplished. Along the way I will discuss some traditional virtues as well, especially integrity, discernment, humility, and courage.

Choosing Projects Well

It matters how we spend our time. One professional with whom I talked considered this a central challenge in bioethics. Asked to name a significant virtue, he offered these thoughts: "Courage; or whatever it takes to be credible; not so much private-life morality, but things like passion, conviction. . . . On the other hand, there are missionaries in the pejorative sense, too, imposing their values on others. Fine line. But [we] should be proactive, not just responding to the questions others put to [us]. . . . The real question is, 'How do you do ethics?' And the answer is, 'From the big picture, from the point of view of the marginalized, from the fact that we exploit people.'"

I think that he is right about the way we should be choosing our projects, and right about the fact that it takes courage. It also demands self-awareness. One of my interviewees was asked to do a series of talks to a church group, whose pastor had taken a bioethics course elsewhere. There were to be three sessions under the general title "When Someone in Your Family Is Ill, . . . Aging, or . . . Dying." The pastor wanted the section on family illness to deal with anencephalic infants, an issue that had fascinated his teacher and so fascinated him. But the woman with whom I spoke thought a discussion of chronic illness—cancer or diabetes—would much better meet the needs of that audience. She believed the pastor had been ill-served by a bioethics course that left him insensitive to the needs of his congregation. Someone else remarked, "I sat on a committee for *ten years* without knowing that pain was undertreated. This is a Catholic hospital—it shouldn't be a problem. But no one ever told us that hospital ethics committees should pay attention to it."

The forces that shape our professional agendas deserve attention. During the week of November 7, 1994, I was struck by how much expensive professional time went into the discussion of Mrs. Bactri's refusal of a gastrostomy. I came to realize that what was at stake was not how much longer she would live, but how decisions made at the end of her life would shape its meaning. I still remember that session as rewarding, in spite of my doubts in principle about ethics consultations. Upholding Mrs. Bactri's refusal of treatment was an honorable and significant activity, and still more so if we contributed to a pattern of respect for patient choice within the hospital.

But it would be hard to argue that this was the only, or the most significant, thing that any of us could have done that day. As I realized toward the end of the week, improving decision making within one area of medicine may divert attention from others. Attention is a scarce commodity, and it does not automatically go to what matters most. We know, for instance, that public concern with drugs, crime, and other issues varies not with the current magnitude of the problem but with the degree of attention the media have given it. That is one of the reasons our interactions with the press are important. But we are catalysts of attention in everything else we do as well, in a private ethics consultation or on a web site visited by millions, in the classroom or presenting Grand Rounds. I don't mean, of course, that we are constantly followed by avid fans. Many ethics committees within hospitals are virtually invisible. I simply mean that, since part of our work is getting people to pay attention and think, we need to be self-critical about where we direct our energies.

Choosing well in this regard may be thought of as a virtue. And to choose well, we must have a rather sophisticated understanding of not only what is important within our domain, but also the forces that move us toward certain questions, and the feasibility of trying to resist those forces. (Audiences cannot just be told, "This matters. Pay attention.") This broad understanding demands a disciplined, habitual attention and openness. In addition, choosing well demands courage. Asking new questions can put one at risk, not personally, but professionally. It can demand self-sacrifice: ignoring the possibility of grant money, or surrendering old easy ways of captivating an audience (whether undergraduates or the dean of the medical school). If "choosing projects well" involves moral understanding, a shaping of emotion and desire, and habitual action, then it's a reasonable candidate for a virtue.

The virtue would demand, among other things, an understanding of the currents in the ocean in which we swim. Why do we find ourselves in some parts of the sea and not in others? It is natural to assume that we are moved by the intrinsic gravity of certain questions—those involving death, in particular. Casting some doubt upon that assumption is an important first step.

Focus: Death

For many years the medical school ethics curricula at MSU focused almost exclusively on death (or what is referred to within the field as "end-of-life" issues). We were not alone among medical schools in that preoccupation. Brain death, persistent vegetative state, the right of competent patients to refuse life-saving treatment, euthanasia, physician-assisted suicide, treating or not treating severely compromised newborns: ten weeks of this, ten different facets of the same fundamental question—when is it morally permissible to take a course of action whose foreseeable result is death?

The question is hardly unimportant. Nor is it a single, simple question: there are many "end-of-life" issues, each raising separate, deep questions. Respecting a refusal of treatment involves questions of patient rights and of what counts as decision-making capacity. Parents' refusal of treatment for a newborn raises very different issues, about their rights over their children and the moral status of infants. And so on; one could spend a productive year-long graduate seminar on any of these issues.

But they are not the only ethical questions that arise in health care, nor are they the most common. Life is saturated with moral questions, assumptions, and challenges, within health care as well as outside it. If morality is a question of properly valuing what is and what could be, then everything we do has a moral dimension. Fortunately that dimension does not need constant attention. We can and must take a great deal for granted, morally as well as practically, but there is endless material for reflection, should anyone look for it. Later I will talk in detail about moral problems in health care that get relatively little bioethics attention. For the moment a few initial examples might help: What does a nurse in intensive care say to telephone inquiries from patients' friends, relatives, neighbors? What kind of hospital advertising is legitimate? How much time should doctors devote to charity care? How, and how often, should alcoholics be treated? Should lab technicians unionize? Should Medicaid patients be shifted to managed care?

One natural response to this list is to say, "Yes. These things matter. But nothing matters as much as death." In one sense, of course, that is true; death is uniquely important. But it does not follow that every *decision* about death is uniquely important. Some decisions to allow death really are profound, but others are not, and they do not exhaust the domain of what matters. Where years of life are at stake (especially years of conscious life), choosing to allow death is a grave matter. Most of the cases that have reached the courts or served as the impetus for legislation—and thus shaped bioethics—concerned this kind of decision. Karen Ann Quinlan and Nancy Cruzan were in persistent vegetative states, and could have lived for years that way. (In fact, Quinlan did; it turned out that she did not need the ventilator support whose withdrawal the court allowed.) "Baby Doe" in Bloomington, Indiana, who had Down's syndrome and was allowed to die from a correctable intestinal blockage, would probably have lived well into adulthood. Dax Cowart, the badly burned man treated against his will, is still alive decades later.

Most questions about allowing death are different. Death today is usually "managed": that is, medical care delays it somewhat, and then discontinuing care allows it. In my experience, ethics consultations are about this kind of decision more often than any other. In one extreme instance a consultation was requested about an unconscious patient with no family, who would have lived a few weeks if given aggressive care, and a few days without it, in neither case regaining consciousness. In another case a baby would have died within a few days with one kind of care, within a few hours without it. Often what was at stake in these cases was the question of who had the right, or should be forced to shoulder the responsibility, of deciding whether to try to extend a severely compromised life for a short time.

This question is not an inevitable focus for bioethics attention. It is too easy to imagine, and in fact to remember, a system in which these decisions were routinized: some decisions automatically fell to doctors, others to patient or family. Even today, many other kinds of decisions are essentially made for patients: "This is what we're going to do." "The leg has to come off." "I'm going to change your medication." (For some of these actions, the patient will be asked to consent, but only as a legal formality. A really resistant patient could say no, but that is generally seen as a practical or psychological problem, not an ethics issue.) I am neither endorsing nor criticizing what is routinely done, but am simply pointing out that it is not the intrinsic gravity of an outcome, but the attention we have learned to pay to *choices* about it, that creates a bioethics frame. Bioethics is often not responding to something naturally compelling, but to something it has helped construct as compelling. That this has happened may be right and good, or it may not; but it is not inevitable.

Factors That Direct Our Attention

If a focus on (choices about) death is not inevitable, then probably no focus is. What are the factors that shape our choice of projects? If we are to choose them well, we need to understand what there is to resist, and what to endorse.

The Impact of History

What we call bioethics arose in the 1960s and 1970s in tandem with movements for civil and consumer rights; medical paternalism and research on unconsenting subjects were seen as violating the same basic rights to information, self-determination, and choice. About the same time new technologies—kidney dialysis, organ transplants, in vitro fertilization, intensive-care units—were forcing hard choices. Within this context novel litigation occurred: Severely compromised lives, for instance, could now be prolonged indefinitely, and sometimes these patients or their families went to court to have life-sustaining treatment discontinued. This history of bioethics has been traced by others in detail, and like any history it includes, besides these broad cultural forces, personalities, power struggles, and historical accidents.[7] For the first few decades the story was largely an American one, and as the field develops today on other continents it retains an American imprint.

The results of those early activities still inform the field: national commissions produced reports which remain authoritative;[8] judicial decisions set precedents;[9] federal regulations about research on human and animal subjects still have as their core what was formulated then. A body of scholarly literature set an intellectual agenda; hospital ethics committees and the custom of consultations about patient care began; undergraduate courses were introduced, attracting from the beginning large enrollments (which every philosophy department needed, and still needs). All of these undertakings helped the field continue, and they continue to shape it.

And so a certain set of issues remains central today: "patients' rights," which is most strongly a right to refuse treatment; "informed consent" by subjects of scientific research; new technologies and the questions they pose, currently those arising from genetic science. I would not argue that we are free to ignore these historical roots. Those who in various senses employ us expect us to address such issues, and their expectations create both moral demands and practical limits. But just as the Supreme Court finds within its own precedents the seeds of different decisions for different times, we need to look within our traditional concerns for fundamental principles that will allow us greater scope.

Economic Forces

Another pattern directing the attention of bioethics is economic; we attend to issues that interest those who, in a variety of ways, pay us. The most basic form of payment is a salary and a secure job. My own center began, in part, because its early nonphysician faculty were willing to work on a temporary, part-time basis. (Nonacademic readers, or academics who have always had full-time jobs, may have no idea how poorly part-time faculty in the humanities are paid. It is nothing like pro rata. It can be less than unskilled labor makes at the same institution; and of course it carries no benefits.) In such a context, how do full-time jobs develop? By convincing the medical school that one's work is important; and the best way to do that is to look at questions that trouble physicians, and to look at them from their point of view. If, on the other hand, one is hired into a college of arts and letters because courses in bioethics are in great demand, we are well advised to cover the material that students find compelling. None of this is bad; the vaunted intellectual freedom provided by tenure is always tempered by such forces. But one result is that we tend to put our energies into what has *already* gripped our audiences.

Another almost equally important economic force is the need to get funding from external agencies: foundations, government agencies, and industry. Attention is now being paid to the way funding from industry affects what gets published, and therefore the information available to other scientists: such funding affects the way questions are framed, data are interpreted, and results chosen for publication. Not a lot of attention goes yet to the way funding determines which questions are asked. When the Human Genome Project began, one of the biggest projects in biological science ever launched, 5 percent of its budget was earmarked for the study of "Ethical, Legal, and Social Implications" (ELSI). This was the largest sum of money ever made available in bioethics, and immediately we began to pursue it. Repeatedly, when I asked my interviewees with "genethics" projects how they got started, the answer was, essentially, "Someone told us about a big pot of money up in Washington, D.C. We decided to get some." This should not be written off as opportunism or venality; many bioethicists either depend on grant money for their own salaries, or work within academic institutions where grant getting is virtually demanded.

The ways in which external funding shapes academic effort need considerably more attention. The picture is complex. To begin with, most scientific research is expensive. It needs high-priced equipment (labs, computers, materials) and a sizable support staff. No university can support such projects on its own. In the United States, the largest single source is federal funding, especially from the National Science Foundation (NSF) and the National Institutes of Health (NIH), but private philanthropy can also be significant, especially for the social sciences. In addition, industry helps fund projects with commercial promise. Each of these sources is influenced by a variety of factors. Federal funds are targeted in accord with congressional mandate; with initiatives established by agency staff; and, since decisions about particular projects are made by panels of nongovernmental researchers, with the community of science. When a scholarly field is in its infancy, the staff of an agency have more leeway; one of my interviewees from engineering ethics described the beginnings of this field as partly an initiative from NSF staff, who were much less constrained than counterparts working with, say, the history or philosophy of science, where fields and boundaries were firmly established.

Private philanthropy is shaped by the vision of their founders and, again, by staff. (The Bill and Melinda Gates Foundation is now the largest in the world. Gates is discovering that giving money away usefully takes as much time, energy, and creativity as earning it. His foundation is run by a former Microsoft executive and by Gates's father; some critics have advised him to trade them for more experienced people from older foundations.)[10] One of the people with whom I talked described how a private foundation worried about overpopulation: "They tried to move the conversation from 'population crisis' to 'global stewardship,' then realized the move would be a 'conversion experience'—and those are religious terms. . . . [T]he conversion experience could be facilitated by churches. . . . But they had failed to get religious groups on board from the beginning, and recognized that they needed help." The result was substantial grant money made available to a faith-based institution.

This intricate web of forces would seem to have little effect in bioethics, which does not generally require labs, equipment, and staff. But the picture for science influences that for bioethics. Medical schools need money and want prestige, and both come in significant measure from getting grants to do scientific research. The result can be pressure on bioethics faculty to get grants, sometimes quite unpleasant pressure. Someone who had left a medical school for a more traditional academic campus remembered how glad he had been to leave. "The exact same day I accepted this job my [previous] dean had raked me over the coals for not bringing in enough grant money. 'Where's the good-faith effort, friend?'"

Grant funding, in fact, is highly prized in any department on campus, because grants usually include an item called "indirect costs"—money to compensate the institution for its overhead. This can be a high fraction of the money awarded to the project. If, for instance, half a million dollars is awarded for some project, the grant will probably carry *another $200,000* to be given directly to the school or hospital, to be put into its operating fund and used in whatever way it chooses. Those who bring in such amounts are prized not only for the research they do, but for the way in which they help the institution survive. Bioethicists working in a medical school will seem like parasites unless they, too, bring in money. This sense is intensified by the facts that most of us are not clinicians (and so cannot bring in patient-care dollars) and teach relatively little. A medical school might require its faculty to spend about 35 hours a year in the classroom, often with a group of 8 to 10 students; their counterparts in philosophy or literature will spend about 200 hours a year, in the classroom with 10 to 300 students. Finally, the few bioethicists who work in self-supporting centers may depend completely on funding from outside sources—sometimes from consulting fees, but almost always from grant agencies as well.

The pressure to bring in grant money, then, is understandable. But it has serious effects on how we spend our time. One person remembered unhappily, "I suddenly found that I was committed to spend the next three summers working on informed consent for tonsillectomies—because that was what we had grant money for." On her own she would have chosen to work on questions she found more substantive. The pressure to pursue grant funding presents steady moral challenges: should this person, for instance, have withdrawn from the project she found trivial? In the next chapter I will look more directly at such issues.

Money speaks in a more minor but sometimes cumulatively important way through speaker fees; some of us earn substantial amounts for speaking to lay and professional audiences. A popular and effective speaker, in fact, can make a great deal of money. Audiences are eager to hear about the dazzling issues, once surrogate motherhood, then physician-assisted suicide, most recently cloning.

Finally, there are metaphorical payments. Journalists do not pay us, but media coverage is flattering. Church and civic groups usually offer only modest stipends at best, but reward us with gratitude and attention. The pressure here, again, is to come up with exciting cases and issues, whether or not they are particularly important.

In many ways these audiences, from church groups and students through medical schools and funding agencies, with their varying powers to compensate us, converge in their interests: each is likely to see the interesting questions in health care as those faced by doctors or raised by scientists. Students and other audiences quite naturally like exciting issues; referees are likely to be conservative, to keep within a field's established boundaries. And so we are reinforced in our attention to some topics and not to others. The virtue of choosing projects well demands recognizing the power of these rewards and considering whether and how to resist them.

Academic Factors

"Garbage is garbage, but the history of garbage is scholarship." I don't remember who made this point, but it's true, and not as cynical as it sounds. Furthermore, once a history of garbage is written, there are likely to be responses, criticisms, responses to criticisms, criticisms of responses. Attention begets attention; scholarship generates scholarship. Soon there's a whole field. "Orphan" pieces, unrelated to what others have been talking and writing about, have a hard time finding an audience.

The rhetoric of scholarly fields also perpetuates certain emphases. One example is the conventional point of view in descriptions of ethics cases. Tod Chambers has pointed out one typical way of reporting the cases that are a mainstay of bioethics: the reports "do not tell the patient's story, nor do they tell . . . the ethicist's story; instead they tell the physician's story."[11] Frequently a case study suggests "the clinical gaze. First it identifies [the patient] by his initials," age, and sex; uses phrases like "admitted to" one problem, "denied" another, "reported" another. Such sentences, written in a passive third person, are nevertheless not neutral. The patient could not tell the story that way. A patient who said "I admitted to no sleep disturbance" or "I reported no childhood diseases" would be suggesting something quite different from what is intended by the conventional "he admitted/denied/reported."

So case studies, whatever their grammatical person and voice, nevertheless present the world from a specific perspective, usually that of the doctor or nurse. In addition there are code words that encapsulate whole ethical discussions: "informed consent" suggests not a process of shared decision making but a patient who listens and says yes; "maternal-fetal conflict" frames a conflict between a pregnant woman and her obstetrician as a choice the latter must make between respecting the woman and protecting her fetus. "Confidentiality" ignores a vast web of questions about who should know what, focusing narrowly on one: whether a health-care provider may reveal patient information to a third party. These technical phrases, these code words, are the tools of our trade. Bibliographies and search engines are arranged around them. Even my own file cabinets are organized accordingly. I have "oncology" but not cancer, "newborns" rather than infants (because bioethics focuses on the questions that are evident in the first days of life, not those that arise during, say, the first year of life). In fact, the only reason I still have a file called newborns is that "neonates" is spelled so similarly that in looking for the latter, I find the former. I do have both "aging" and "geriatrics," however.

The virtue of choosing projects well requires noticing such things, holding on to labels like "childbirth" rather than "obstetrics," habitually asking whether what is included within standard labels could be expanded, and struggling to widen or divert the happenstance currents of scholarly fashion.

Through a Doctor's Eyes

I asked someone who works broadly in professional ethics why he takes up the topics he does. "They're what bother practitioners. Truth telling in the courtroom bothers lawyers. . . . Engineers are [haunted] by stories of buildings that fall down, deaths and injuries that follow." And so are doctors, of course, haunted by death.

In addition, in a doctor's daily professional life, as one commented, "Refusal of treatment really gets our attention." Such refusals are disruptive rather than haunting, although the refusal of life-sustaining treatment is both, and this fact goes a long way to explain bioethics' focus on death. Arguably a first responsibility of anyone in professional ethics is to do exactly this, deal with what is most troubling to members of the profession. If it's first, however, it is not also last. There is an instructive contrast in the academic field of criminal justice. Researchers, investigating the extreme unreliability of eyewitness identifications, developed methods of conducting police line-ups that substantially reduce the number of misidentifications, *even though lawyers remain uninterested in the results and make no use of them.*[12] It helps that these researchers were tenured academics, not dependent on lawyers for their living or their status.

In bioethics most of us are not so independent, and this makes it harder for us to do what someone whom I quoted earlier demands of himself: to remember the big picture, the perspectives of the less powerful. Take, for example, the fact that doctors and nurses can make problems for one another; either can interfere with good patient care or be personally difficult. But it is only nurses who identify such problems as ethical issues. Their ethics textbooks often have a section on doctor-nurse conflicts.[13] I've never seen a textbook, or even a syllabus, for medical students that mentions the issue. The reason is that the difficulties are not symmetrical: nurses have real power but little authority, while doctors have both. When nurses get in the way of doctors' doing their jobs, the problem is largely seen as practical. Can the doctor make the nurse see reason, or remove him from the case, or should she just put up with the unpleasantness? For nurses, however, the problem can be a moral one. Because they have less authority to work on their own, they may feel forced to take part in care they believe is inappropriate, dangerous, or ethically wrong. Finding a way not to do that takes both courage and discernment. The asymmetry in power leads to an asymmetry in moral situation, which few doctors recognize. (Good doctors do recognize that good nursing care is essential, and that competent nurses may know things physicians do not. These physicians routinely consult and collaborate with nurses. It's the further step, the understanding that doctors can put other health-care workers in a moral bind, that is usually overlooked.)

Furthermore, most doctors' professional lives are spent caring for patients one by one. (How else could patients be cared for?)

So it's not surprising that what troubles doctors tends to be decisions about individual cases. When they pay attention to the ethical aspects of organizations and systems, it is ordinarily because something interferes with the care of individual patients, or because the doctors believe they themselves are being mistreated in some way. The massive attention that went toward issues in managed care in the middle 1990s arose first because *doctors* were frustrated. They had been trained to do everything possible for their patients as individuals, not to consider whether using money for this patient would end up hurting others. The public reaction to managed care began slightly later, but it took much the same course: middle-class "paytients" (as Len Fleck puts it) were not accustomed to restrictions and limitations. Of course some of the changes brought by managed care were unwise and unethical, even if some were necessary and fair. My point is that attention to systems entered bioethics at the point when doctors began to find systems seriously troublesome. Nurses have always worked with limited resources, always had to practice within unwieldy institutions, always had to balance the patient in one room against those in the others. Those problems never got the attention of bioethicists. One person (not a nurse) with whom I talked said vehemently: "Put this in your book: Bioethics apes power! It follows doctors, not nurses." That is a significant reason these questions are not prominent on our agenda.

With a Philosopher's Tools

One more factor sometimes focuses bioethics' attention: the limitations on what philosophical skills can do. Not all bioethicists are philosophers, but many are, and some aspects of the discipline have been influential. Modern philosophical ethics, to the extent that it directly engages action, tends to do so at two levels: individual actions, or the shape of society as a whole. For centuries it was blind to intermediate structures—families, communities, institutions—and so has few tools for understanding and evaluating them.

I don't want to paint too simple a picture. Despite the way economic, institutional, and disciplinary forces converge, other factors sometimes counter them. Recently patient-rights groups (especially

AIDS activists) forced attention to the right to be included in clinical trials, not just the right to abstain. Currently serious attention is being directed to the conflicts of interest that arise when biotechnology invests in scientific research. It's probably also true that the appeal of Doctors' Dazzling Dilemmas is waning, as fresher issues summon bioethics attention. But Science Sensations will always be with us.

Some Topics We Ignore

One of my interviewees believed that "we drift like a tide with whatever is hot." She illustrated her point by noting that national health care was virtually abandoned as a topic after the Clinton proposal failed. That is one of many examples of inquiries not launched, or abandoned too soon. To appreciate the sometimes peculiar boundaries of bioethics, it helps to think about what is outside them.

The Uninsured and the Poorly Insured

Roughly 45 million Americans have no health insurance—neither private insurance nor Medicaid nor Medicare. Because Medicaid reimburses poorly, the additional tens of millions it covers often have great difficulty finding a doctor or hospital who will treat them. In one town in Michigan, no obstetrician will accept Medicaid patients; women in labor must drive fifty miles to the doctor they've been seeing, and a country hospital en route describes itself as the place where women stop when they can't make the whole distance. In big cities the concentration of the uninsured and the poorly insured can make the hospitals who treat them insolvent; the hospitals close, the burden is shifted, other hospitals farther away take it up for a while—but for the sick poor, "farther away" can be unknowable and unreachable.

Bioethics has consistently paid some attention to the uninsured. Textbooks for undergraduate bioethics courses include units on "justice," focused on the question of whether there should be national health care or insurance. Philosophy has an extensive literature on questions of distributive justice and most philosopher ethicists, although not all, concluded long ago that health care should be a right rather than a consumer good. But once that conclusion has been reached—and here is the point—philosophy has little more to say. The subject does not usually come up in medical school courses, nor in workshops or presentations to health-care audiences. The poorly insured get even less attention.

Things were different for a brief period of time in the early 1990s, when health care became a significant issue in electoral politics and Bill Clinton proposed a form of national health care. Bioethicists at that time said and wrote a great deal about the subject, and some took part in the task forces that drew up the Clinton plan. Afterward, those I know dropped the subject for years. Partly this was because audience attention had gone elsewhere; it was about that time that managed care became such a painful issue and the human genome project was breaking ground. But I believe that a sense of hopelessness, even heartbreak, also contributed to the silence. Some of my closest colleagues had worked a lifetime toward universal health care, and had seen the best chance of a generation fail.

The plight of those with little or no health insurance, then, fails the tests of what is likely to get bioethics attention: it is not a problem faced by most doctors in their daily work; there is nothing new or exciting about it; there is no public debate creating an audience for discussions of it; philosophical ethics sheds little light on its cause or on the practical pursuit of a cure. I would have hoped that at least the question of charity care would arise—do doctors and nurses have a moral obligation to spend part of their time giving uncompensated care? I broach the question, at least, in my work with third-year students, and offer material on it to the faculty teaching them on our community campuses around the state. Neither students nor faculty express interest.

Social Determinants of Health

Still less frequently mentioned is another fact of great philosophical and moral interest, but so little subject to individual action or even to legislation that it falls outside every bioethics category. That is the profound influence of socioeconomic status on health. The influence goes far beyond the obvious fact that poverty correlates with poor health. Audiences are never surprised to hear this, and generally attribute it to lack of health care and to bad health habits. (The first is often true; the second may or may not be.) Two other major contributors are virtually ignored, I believe because the American individualist lens is so powerful. That is, we tend to explain everything in terms of characteristics of individuals, and to be blind to the fact that systems have properties that individuals within them do not. (The heart, for instance, can be misshapen even if each cell in it is healthy. Bodily tissues have color although their constituent atoms do not. And so on.)

The first overlooked factor in the health status of the poor is the nature of poverty itself, which makes it hard to deal with even the ordinary details of daily life. Money is correlated with, indeed essentially *is,* the ability to control the circumstances of one's life. The poor must worry every day about whether they can keep warm, pay the rent, feed their children, and get to work. Not much on this list is reliably within their control, and a sense of control has frequently been correlated with health status.[14]

What we know about heart disease illustrates this point well. Feelings of hopelessness, despair, and failure correlate strongly with the risk of death from heart disease. Some studies show that these factors are considerably more significant than better-known risk factors like raised cholesterol levels and hypertension. In the American context these conclusions are all too likely to be translated into facts about individual psychology. The terms "despair" and "depression" are easily interchanged, and when that is done, certain mantras arise automatically ("depression is a treatable disease") and certain causal explanations begin to be assumed: it's a chemical imbalance in the brain; it's a result of your genes; it's the way you were raised.

But it might also be a result of your social situation. Some people really do have less control over their lives, and less to hope for. One study linking hopelessness with the risk of heart disease found "depressed affect and hopelessness . . . more common among women, blacks, and persons who were less educated, unmarried, smokers, or physically inactive."[15] Note the way the sentence structure suggests that these factors are all the same kind of thing, all properties of individuals rather than of relationships among individuals. But gender, race, and marital status are essentially relational, as is "*less* educated."

Even emotions (like hopelessness and despair) are intrinsically more relational than we usually recognize. As one analyst notes, emotions "are more than psychological states. They are also reflections of culture and social structure. This is why, for example, anorexia nervosa, a form of self-imposed starvation [accompanied by certain emotions and beliefs], is found almost exclusively among teenage girls and young adult women in western industrialized societies, while *taijin kyofusho,* a morbid dread of causing embarrassment to others, is observed primarily in Japanese society. Feelings of despondency, depression, and despair are now understood to be so culture specific that guidelines for psychiatric evaluation recommend that cultural background be considered in diagnosing [them]."[16] Similarly, culture will determine what counts as success and failure, and powerfully shape what beliefs about the future count as hopeful or hopeless.

There is much more to understand about the health status of the poor than whether they have access to doctors and whether they eat wisely and exercise. One way to reinforce this point is to note that in the United Kingdom, where health care is universally available, and where health statistics are collected by socioeconomic class, there has been a constant correlation between class and health. It's not just that the poor do badly. Those at each of the five steps do better than those one step below them. It's a steady, staircase descent.

Finally, there is a great deal of recent evidence showing that the degree of inequality within a society *in itself* contributes to poor health, and equality to good health. Of two societies with the same median standard of living, that with less difference between rich and poor will have the better average health. The affluent in a society with extreme inequalities will be less healthy than the affluent in a more egalitarian society—as will the middle classes and the poor compared to their counterparts. This leads to surprising, even ironic, conclusions: for instance, the best way to improve the health of the rich may be to lessen the degree of poverty.[17] And the correlation helps to explain why Costa Rican health statistics equal or outstrip those of the United States.[18]

As interesting as these facts are, one may wonder whether they have any relevance to bioethics. After all, some stratification is inescapable in societies of any size, and the extreme stratification within the United States is not the result of anything doctors, nurses, or patients do. My claim, however, is that part of our job is to improve the moral perception of our various audiences. If Americans tend to be blind to one of the most morally important characteristics of our society, and a characteristic that sharply impinges on health, then bioethics should at the very least be helping us see.

Hospital Workers

I have already mentioned the fact that nurses can find themselves morally compromised. The same is true of dietitians, social workers, and anyone whose work with patients is subject to the authority of others. When the problem arises about the care of an individual patient (say, tube feeding for a patient who seems neither to want it nor to benefit from it), mainstream bioethics offers some help. An ethics consultation can be called, and there is a significant body of literature addressing the question.

For other situations, ethics consultations offer no help. Think of a chaplain, told by a hospital vice president to make a "home visit" to a recently discharged patient, a state legislator. The chaplain believes this is unjust, that his limited time should be spent helping ordinary patients still in the hospital. But he has little choice in the matter. Or think of nurses on units where tasks once considered part of nursing have been reassigned to housekeepers and orderlies. Such "restructuring" was common in the late 1990s, since nursing salaries are a large part of any hospital's budget, and since some traditional nursing tasks look unskilled: changing catheter bags, recording food consumption, moving patients, and so on. These reasons, in addition to the severe shortage of nurses now looming, lead hospitals to staffing patterns—"restructuring," understaffing, mandatory overtime for exhausted personnel—that nurses believe shortchange and endanger patients. One nurse in such circumstances said, "Every day I leave [the hospital] feeling that my heart has been torn out."

Such nurses suffer from what is called moral distress, a different phenomenon from the dilemmas to which bioethics typically responds. Dilemmas create puzzlement; they arise when it is not clear what is the right thing to do. In contrast, moral distress is a sense of complicity; it is not uncertainty about what is right but the belief that one is virtually forced to take part in what is wrong. Moral distress is likely whenever one human being is subject to the authority of others; it is not a symptom of evil but a sign of the human condition. We are social beings and we are moral agents: because we are social beings, most of what we accomplish depends on the coordination that institutions (with their hierarchies) make possible; on the other hand, because we are moral agents, we can never simply agree to follow orders. Bioethics offers little on this topic. A few pioneering writers in the 1980s named the phenomenon, but little further work has been done; and the field offers nothing in the way of practical help.[19]

These are not the only interesting ethical questions concerning hospital employees. They can be exploited, for instance; and conversely, they can use their own power unfairly. Should lab technicians strike? Should employers try to block or break unions? Should nurses bargain for twelve-hour shifts, knowing that these schedules are not ideal for patient care? (Most nurses are women; women are likely to do most of their families' child care and housework. Because of this "double day," borne by women around the world, American nurses sometimes press for, say, three 12-hour shifts a week rather than five eight-hour ones.) When exhaustion sets in toward the end of the day, patient care may be compromised; furthermore, scheduling difficulties become formidable. When some are on twelve-hour shifts, some on eight, some possibly on four, some working three days a week, or four, or five—under these conditions, real teams, groups who know one another and work together smoothly and efficiently, cannot emerge. If bioethics has little to say about how hospitals should treat workers, it has still less to say about how the workers should treat the institution.

Nor do we have much to say about how health care professionals should treat one another. One nurse manager described a pledge he periodically asked his staff to take: "Some people find it hokey, especially 'I will not engage in the "3Bs" [Bickering, Backbiting and Bitching].' And it's not a magic bullet. It just puts into words basic decent human behavior. But how often do we do that? Once in a while I'll suggest, in general, the need for forgiveness; but that's a higher-level ability, not everyone can manage it, and I never ask it of individuals."

He has a moral goal, to help his staff work together in mutual respect; and he understands the difference between basic decency and higher ideals. He's engaged, in other words, in ethics work. The tools he uses emphasize talking with those with whom one has trouble. When I talked with him he had just concluded the yearly performance reviews for his floor, in which he discussed with everyone, and had each once again sign, a "Commitment to My Coworkers" statement. "I've felt a real lightening of our spirits since then; energy has been freed up that can go to patients. That's the way I sell this to my staff." His unit also had remarkably low turnover. Low turnover means experienced staff, experience brings with it efficiency, and the result is that nurses have more time to review the plan of care and more confidence in approaching the medical staff with their concerns.

Patients benefit when there is mutual respect among the staff. But this nurse manager knew that mutual respect is also a moral good in itself. His help in understanding and fostering it came from management literature, not from bioethics. The electronic search engine Bioethicsline in June 2000 listed over seven thousand articles on confidentiality, only two of which were about gossip.

The Patients' Perspective

One young physician gave me his personal "genealogy" of bioethics: "First there were the precursors, like Joseph Fletcher [theologian author of *Situation Ethics*] and Henry Beecher [physician who blew the whistle on medical research abuses]. Then there was the first generation, people like Dan Callahan [founder of the Hastings Center], and next the generation trained by those pioneers. Now . . ." "Now?" I asked. "Now there are people like me. Doctors."

It seemed clear to him that the field, at least in its clinical aspects, had finally reached its maturity. The proper people were at last in charge. Nonclinicians had been indispensable as pioneers, but now the torch had passed. (If pressed, I believe he would have included nurses, therapists, and others from the bedside as acceptable in a mature clinical bioethics.)

A nurse from the same institution agreed with this description, but not with his endorsement of it. Her remark, to the contrary, was that "the patients' perspective is dropping out of bioethics." If, as I argued earlier, bioethics tends to see issues through the eyes of physicians, her remark is not surprising. Note that she didn't say patient *issues* are dropping out of bioethics, but that their perspective was. She went on with an example: "Most of the debate about assisted suicide is about what it's proper for a doctor to do. It's not about what patients need and want. Or deserve." Once again, my own files attest to that: they contain a folder called "Physician-Assisted Suicide" and carry articles about Jack Kevorkian, about the Oregon law and the Supreme Court's response, and articles by ethicists commenting on all this. I have no folder labeled simply "suicide" or even "assisted suicide."

The perspectives of doctors, nurses, and therapists are essential in bioethics. They understand things no one else can, and can speak to one another with force and candor. But their perspectives are not enough; some important concerns are only visible from other angles. Take, for example, the question of assisted suicide. Margaret Battin's "Assisted Suicide: Can We Learn from Germany?" describes a cultural and legal context very different from that in the United States.[20] American attention has generally gone to Holland and not to Germany; in Holland, physicians may and sometimes do perform euthanasia. Germany deals with the issue quite differently: as a result of its Nazi past, it rejects euthanasia out of hand, but suicide is not illegal, nor is assisting it. It is not doctors, however, who provide the assistance. Suicide among the very ill does happen, not uncommonly, but it is almost always a private action rather than a medical one. There is a national society dedicated to helping those who want to end their lives, something like the Hemlock Society here.

Battin's article suggests a number of moral and ethical questions, including ways of understanding suicide and the proper role of private organizations formed around the issue. She is careful to talk about the subtleties and difficulties of trying to learn from other countries. She frames the article, however, within the pressing practical legal question of whether physicians in the United States should be legally empowered to assist with suicide. This is natural in many ways: the legal question being debated was about doctors, not about private organizations, and bioethics has always been driven by practical and especially by legal questions. In another way, however, her emphasis is odd. Because the configuration of law and custom within German society is almost impossible to imagine here, the more useful focus of her discussion would have been the proper roles of "aid in dying" societies. About this issue little has been written. Nor is there much about the moral nature of suicide, its justification or phenomenology. The question so heatedly and thoroughly addressed is "What should doctors do?" This discussion offers no help to the patient who might be trying to think through her situation, nor to the layperson to whom she might turn for help.

Another example is that physicians can be oblivious to the fact of, and the moral relevance of, health-care *dis*organization. I am not speaking here of insurance companies and managed care, both the target of constant attack. Instead, I am talking about the kind of problem that threatens any complex organization. The movie *The Doctor* illustrates this beautifully. The central character, a physician with cancer, is shuttled from one office to another, one clinic to another, until he finally erupts: "Is there a rule in this hospital that everyone who deals with a patient has to begin by saying, 'I'm sorry, but . . .' 'I'm sorry but you'll have to give us this information again, on this other form . . . I'm sorry, but your records have been lost . . . I'm sorry, but there was a misunderstanding about the time of your appointment. . . .'" So many doctors are involved in the character's case (and again, this is typical) that no one is quite sure what he has been told. As a result, he learns bad news inadvertently: "As you know, your cancer is growing. . . ." One physician ethicist with whom I work in Lansing frequently remarks, "What this patient needs most is a doctor"—meaning someone who will unify the fragmented care the patient has been receiving from rotating teams and crowds of consultants. (In a teaching hospital students and young doctors move to new units every few weeks.)

One of the most basic concepts in bioethics, that of suffering, should tell us that these facts have moral relevance. Suffering is not the same as pain and injury; instead, it is a way of experiencing them, which follows closely from the meaning one attaches to them. Eric J. Cassell construes suffering as the sense that one's identity is threatened, that disintegration looms, often beginning with the inability to accomplish those things one considers important. Cassell notes that "even in the best settings and with the best physicians, it is not uncommon for suffering to occur not only during the course of a disease but also as a result of its treatment."[21] I do not know that he had systems failures in mind; but, in fact, for anyone confronting illness, being able to talk with a doctor, get tests done, find out the results, perhaps choose treatment but certainly have a say in how it is scheduled, are highly important. To be systematically frustrated in them is a kind of suffering, but one that doctors have trouble recognizing. "To be effectively treated," Cassell remarks, "suffering must be recognized, but that is not a simple matter."[22] In fact, as I discuss in the chapter on moral development, it is a moral accomplishment. Research suggests (and any conversation with laypeople will confirm) that dysfunctional systems are common in health care. Doctors are not always aware of these problems, and do not always take them seriously. To some extent this is the result of their training, focused on the deep technical intricacies of biological functioning and on the quality of their personal, individual interactions with patients.

Systems failures, then, and clinicians' overlooking of the failures, are areas where the patient's perspective is important, but too often missing, in bioethics. There is a crucial, unbridgeable difference between what clinicians witness and what patient and family experience. Another example of the way patients' perspectives can be erased is one I've mentioned earlier, situations described as "maternal-fetal conflict." The label expresses the clinician's perspective and erases that of the pregnant woman, for whom the conflict is between herself and the health-care professional.

Other Examples

Many other ethically important topics receive scant bioethics attention. Some are mundane, some not: Under what conditions should one agree to have someone involuntarily committed, given that conditions besides psychiatric illness can underlie disorientation, and that facilities vary widely in quality? What is the most responsible way to reduce and dispose of medical waste?[23] How do reimbursements to oncologists influence their recommendations about chemotherapy? One young oncologist believed the influence is corrupting: "Watch the language with which the chemotherapy is offered. Patients will be told, 'We don't know for sure that this won't work for you.' What they should be told is, 'There is no reason to believe this will help you, and a great deal of evidence that it won't.'"

Interestingly, the questions raised by this young oncologist fit the pattern of classic bioethics issues; that is, they concern decisions made by doctors about the care of individual patients. Perhaps these questions are not taken up because not a lot of doctors find them distressing, and nothing has happened in the public arena to make them seem problematic. Their invisibility speaks to other kinds of questions—the silencing of nurses and young doctors, the ways in which the media accept and perpetuate a certain stance toward cancer.

Practical Steps toward Choosing Well

The virtue of choosing projects well is required if we are to help people see such issues. It requires keeping in mind the field's moral occupational hazard: whenever we choose to focus attention on one issue, we in effect draw it away from others that may be more important. Putting that concern aside, however, even work properly focused on a certain question suffers if the fabric of related questions is not recognized. Bioethics is like a set of stakes pinning down a great billowing canvas. Some areas are so thoroughly anchored that the images on them can be read, evaluated, and corrected if necessary. Other parts of the canvas surge and undulate out of reach, threatening to loosen what has been anchored, and making even that hard to read. Context is important, and some of the habits of bioethics obscure it.

Choosing projects well demands courage, discretion, and something like night vision, the ability to see when the light is dim. Although there is no formula for developing it, there are some rules of thumb. The factors described above provide a useful set of questions with which to begin: we can develop the habit of noting whether we are too bound by the problems of physicians, the dazzle of dilemmas, the lure of funding, and so on.

Listening to the "Other"

Beyond asking ourselves those questions, however, it is also useful to seek out the voices of outsiders and of the young. Thomas Kuhn suggested long ago that those not fully socialized into a field have the most original vision.[24] The story of our medical school curriculum, whose initial focus on death I discussed earlier, is an example of the usefulness of outside perspectives. A few years ago we added to the curriculum some minority-focused issues, including the Tuskegee syphilis study; that quite naturally led to material on clinical trials in general (in the form of questions faced by primary-care physicians). This inclusion was a first step away from our too death-focused curriculum. Attention to those who tend to be excluded, in other words, should not just be seen as an act of justice; it will often broaden our understanding of apparently unrelated issues. Another instance of this broadening occurred within the Joint Commission for Accreditation of Healthcare Organizations (JCAHO), which has enormous power over what hospitals do. Until about thirty years ago, its standards emphasized patient welfare but said nothing about patient rights. (The difference is important when a patient wants something that hospitals believe is not in his or her best interests.) The standards began to change in 1969 when the Joint Commission sent out a revision of its standards to a number of different groups for public comment. This was simply a move toward hearing from a broader set of people. And it was the National Welfare Rights Organization (NWRO)—not doctors, nurses, or hospital administrators—who proposed including patient rights. NWRO was an umbrella organization comprising more than 300 local welfare rights groups, whose members in turn were welfare recipients and welfare activists. Once again, listening to outsiders and minorities led to a broader understanding of one's responsibilities generally.

The National Bioethics Advisory Commission (NBAC) did something similar. Its mandate was to "provide advice and make recommendations" about the "appropriateness" of governmentally sponsored bioethics research; to "identify broad principles" for the conduct of such research.[25] It would be easy to interpret that mandate as calling for the same kind of work one would publish in a bioethics journal, focusing closely on what makes cloning, for instance, morally permissible or impermissible. But NBAC wanted to hear a variety of perspectives, and invited a variety of speakers, including those with theological commitments. This move received a certain amount of criticism: religion is a private matter, maintain the wall of separation between church and state, and so on. But the result of listening to such voices was a significant reconstruction of their job: not to make recommendations about the rightness and wrongness of certain practices, but to make recommendations about *ethically defensible public policy in a deeply divided democracy.* That means paying attention not just to, say, the legitimate use of stem cells, but also to what it means to respect a minority. It means asking about more than the moral status of the zygote; it means asking about the moral distress of those who oppose the proposal, refusing to dissolve that distress into mere preferences; being suspicious if the benefits from a program accrue only to those who approve of it; asking whether objectors would be forced to become complicit, asking whether they would be allowed to use democratic process.[26] Once again, listening to the "other" broadened the moral landscape in unexpected ways.

Even when the conversation is confined to "insiders"—to those explicitly involved in health care, or to those in the majority—it is important to hear from a variety of professions and types of institutions. For one thing, each highlights a different aspect of a problem (biochemistry, social situation, the impact on a sense of meaning); in addition, each includes a set of conventions, of habits and norms, that can be set in illuminating contrast with one another. I began to see confidentiality in a new way, for instance, after hearing nurses and social workers talk together about it. The two fields had quite different customs about how much patient information to share *with one another.* Before, I had thought of confidentiality as it is usually presented in bioethics, as a question of how much patient information to reveal to laypersons, especially to family members and government agencies. Now I began to think about the way information flows within a hospital, and to see that the currents of talk (about patients and about one another) are virtually unimpeded. This awareness brought to mind comments I had heard in other contexts. During a discussion of testing hospital employees for HIV, for instance, someone in the audience had said, "If one of us were found HIV positive at our lab, the news would be all over the hospital within twenty-four hours." Assent swept through the crowd. Having heard that public health departments protect patient confidentiality better than hospitals do, I talked with nurses from that setting about the question. They agreed with the impression I had, gave me examples of the lengths they go to protect patient confidentiality, and said that the stricter norms apply to employee information as well: "We don't worry about information [even about one another] leaving the lab. It won't." I use that anecdote in speaking to doctors and nurses, to show that different customs are at least possible.

For the same reasons, interdisciplinary work is crucial. Each discipline sheds light on a different aspect of life, has its own rigorous tools and its own skepticism. The sociologist sees patterns of power and the detailed structures within institutions, not just the political and economic factors of society as a whole. The anthropologist is able to find the meaning that situations carry for those involved. In a later chapter I will talk in detail about the importance and the challenge of good interdisciplinary work.

Finally, attending to a variety of religious perspectives is enlightening. Hospitals founded by Catholic religious orders, for instance, often had departments called something like "Mission Effectiveness." These departments had a voice in a wide variety of institutional decisions, including mergers and acquisitions as well as care for the poor—a breadth of concern far broader than hospital ethics committees are accustomed to taking up.

Seeking Long and Wide Views

Another useful tack is to take a long view, to track the way a question has appeared over the years. More than one person, for instance, believes that within ethics consultations a pendulum has swung, that, whereas ten years ago the common problem was patients or families wanting less intervention than doctors did, today patients and families tend to want more.[27] In another instance of this kind of shift, I was startled recently to hear medical students talking about women who wanted cesarean sections their doctors did not believe were "medically indicated." The issue used to be just the opposite: women refusing c-sections their doctors felt were imperative. Both shifts (or apparent shifts; I have only anecdotal evidence of either) led me to wonder whether patient preferences trail medical opinion by about ten years, a certain kind of socialization of the public taking place largely through the media so that patients finally come to want what doctors think they should want, just as doctors move on to a new standard of care. The whole picture (of what gives rise to disagreement between doctors and patients) is obviously more complicated. But noticing this curve has made me still more attentive to the power of the media, and more concerned about the ways in which "health promotion" campaigns can go wrong. I would not discuss doctor-patient disagreements without taking this apparent historical trajectory into account. Obviously the deeper and more critical view of professional historians is still more important.[28]

It is also useful to try to see any specific issue as an instance of a broader one. To continue with the example of confidentiality, once I began to see the standard issue (about revealing patient information) as only one of many questions about the way information flows within health care, information not just about patients but also about workers, I came to realize that while the flow is too free within hospital walls, arguably too little information leaves those walls. It may be common knowledge within a medical community that Dr. Y is incompetent, unethical, or even criminal, but patients will be the last to know. The question of who should be told is hard, of course, but important, and like everything else I've mentioned, it gets very little bioethics attention.

Seeing a narrow issue in larger terms may not be something that we can seek directly. Our growing awareness will more likely result from the kinds of listening I've encouraged above. Another example touched on earlier, disagreement between pregnant women and clinicians, bears this out. I began framing the issue differently after hearing from quite a range of women in a variety of settings: sociologist Barbara Katz Rothman, anthropologist Birgitta Jordan, childbirth educator and researcher Libby Bogdan-Lovis, and midwives, some of whom were "direct entry" or "lay."[29]

Being an Outsider

Finally, a number of thoughtful commentators have mentioned the importance of remaining in some sense an outsider oneself.[30] The term "stranger" is often used. The sociological literature cited here contains a number of different terms and models, as does the less frequently cited but still more interesting feminist literature. The early sociologist Georg Simmel remarks that being a stranger entails a certain degree of contact: inhabitants of a distant planet are not strangers unless we meet them. His archetypal stranger is a trader, whose firmest roots are elsewhere but whose interaction in the territory is systematic and ongoing.[31] This trader is freer to leave than the natives with whom he interacts, is outside many of the bounds of convention, and has "objectivity": "a peculiar composition of nearness and remoteness, concern and indifference."[32] The stranger is often chosen as a confidant (a point that triggered for me memories of listening to an anguished neonatologist). Writing somewhat later, sociologist Alfred Schuetz uses a different paradigm for "stranger," that of an immigrant, or any "applicant for membership" in a closed group such as a family, club, or nation. Where Simmel spoke of objectivity, Schuetz speaks of disinterest: the stranger "intentionally refrains from participating in the network of plans, means and end relations, motives and chances, hopes and fears, which the actor within the social world uses for interpreting his experiences of it. . . . [A]s a scientist he tries to observe, describe, and classify. . . . The actor within the social world, however, experiences it primarily as a field of his actual and possible acts and only secondarily as an object of his thinking."[33] Edward A. Tiryakian, still later, remarks, "'Stranger' is not a static category . . . [but one] that any one or any social group may move into and out of, most typically by wandering or migrating. . . . But colonization also produces a different kind of strangers, those who have become strangers in their own land."[34] All this supplies a rich lode of material for thinking about being a stranger, an "outsider within." This last insight is crucial to feminist writers like Patricia Hill Collins, who coined the phrase to describe the situation of African Americans and who points to the epistemological advantages of the situation.[35] Gloria Anzaldua and Maria Lugones employ powerful metaphors to convey the ways in which one's very identity can be mixed, troublesome, enlightening, and useful. (Anzaldua: "This Bridge Called My Back"; Lugones: "curdled identity.")[36]

These approaches cannot be homogenized. Obviously there are many different kinds and degrees of outsiderness. It was important to Simmel's first "stranger," the trader, that he could leave; his primary identity was elsewhere. The feminist writers are talking about quite different situations, situations where leaving is impossible. Yet there is a common thread among the theorists: distance and difference can help one see, but do not necessarily help one be effective. Outsiderness translates into power only when it provides some independence and when insiders *value the perspective* it provides. The young oncologist with whom I talked, who had turned from curative to palliative care and to ethics, was mocked by his former colleagues: "You really like holding patients' hands?" Nurses who learned to speak up, in one study, *never* found that doing so made a difference in patients' care.[37]

Nevertheless it is true that the right kind and degree of outsiderness can be useful. I talked with one consulting bioethicist who was a "stranger" in several senses: He was not a clinician, and not a regular hospital employee, since he was contracted out by his home institution and had other duties elsewhere. The hospital board of directors had approved his hiring and charged him with "being visionary." His official job title was simply "ethicist." That combination of visibility and "outsiderness" meant that people came up to him with a wide variety of concerns—whatever struck them as having to do with ethics. He offered this example: "We include a brochure about ethics in the literature we give patients; it's very general, leaves the word undefined. One day someone came up to me to complain about the food—not exactly an ethics question. I thought. Then I realized it was really a concern about fidelity. He had been promised something, face to face, by a hospital employee."

This man was an outsider in a number of senses: he had the independence of Simmel's trader, as well as the legitimacy. Furthermore, his language was to some degree a pidgin: listeners thought they understood the word "ethicist" but interpreted it more broadly than most "ethicists" do. All this made it possible for him to see and to take up unconventional questions, even those that could be threatening to those with power. He defined ethics much more broadly than has been usual, in part because of the religious tradition from which he drew: he saw employee relations, for instance, as an area in which abuse of power was possible. In addition to recognizing that fact, he also took action, drawing up a questionnaire assessing the "moral climate" in which employees worked.

Choosing projects well, then, if it is not a virtue, is at least a moral excellence, one crucial to our work. It demands understanding the forces that influence the field and the ability to respond intelligently—sometimes resisting them, other times using them as sailors do the wind, to move in a different direction. The virtue demands seeing what is overlooked, a vision that is helped by listening to a variety of voices and by being, in useful ways, an outsider within. But it is not just a matter of seeing; it also involves doing. As one of my interviewees remarked, it calls for a most unscintillating virtue: time management. He gave a mundane example, but one that shows he had thought through his priorities. "I get more speaking invitations than I can accept. I'll choose professional audiences over, say, church groups and the Elks, the kind of groups that have a speaker every month on 'Whatever.'"

It is not just the constraints of money, tradition, and rhetoric that keep us from choosing reflectively and well; we must also confront the lure of internal and external goods. To that issue I turn next.

8 The Goods We Want and the Goods We Need

A Call for Integrity and Discernment

Certain seductions can interfere with doing our jobs well, whether by corrupting our choice of projects or persuading us to remain silent or to allow ourselves to be used in the wrong cause. In the words of one of my interviewees, "There are a lot of ways to sell out. It's essential to have integrity. Otherwise people will sense it and won't in the end listen to you. Some have joined the field out of opportunism, the chance to make a buck. And/or they're too eager to please constituencies, to be pure mediators. Always looking to see how this benefits themselves. People see through that." Someone else made the point more innocently: "A virtue for bioethics is asking what can I give, rather than what can I get."

But selfishness and selling out are not the most interesting issues here. Deeper and subtler questions are found everywhere in bioethics: the need to choose between goods, the fact that one kind of success can interfere with another, the way that apparent good can drive out real good. The danger of "selling out" arises from what MacIntyre calls external goods; for a bioethicist, these include status, money, and public attention. Beyond what we might want for ourselves, there are goods that we need or seem to need in order to do our jobs; among these instrumental goods, funding is the most salient. Finally, there is a good that we need in order to be bioethicists at all, a good that in part defines the practice: we need to be heard—which often translates into a need to please. Contending with the pull of these various goods requires integrity and discernment. By integrity I will mean, as Martin Benjamin puts it, a correspondence between what one believes, says, and does. By discernment I will mean the ability to distinguish the true good from the apparent.[1]

The Lure of External Goods

The simplest problem to talk about is the lure of external goods, "worldly temptations," in another discourse. I saw a minor example years ago: the director of one center was in vigorous pursuit of additional office space for his faculty, most of whom had several offices scattered around the hospital, their appointments with the center being part time. He had convinced himself that each needed a private office in his unit. I've forgotten the rationale, and I don't know whether he was successful. Not much was at stake here, although office space was scarce and the institution might have been better served if the space had gone to others. The situation made an impression on me, however, because the director's motivation was so transparently self-aggrandizement: the center was an alter ego, and its size was part of his status. I doubt that he understood his own motivation, however.

Other situations have been more serious. One person remarked in amusement that he had gotten phone calls from the media within the last month about three different issues, each of them new to him. In each case he gladly responded to the request, confident (apparently) that whatever he said would be of use. It did not occur to him, so far as I could tell, that sending the inquirer to someone else with more knowledge would better serve the public. I don't know the reason for his insouciance, but one possibility is that, for some people, being quoted by the press is quite gratifying.

In my observation, self-seeking, whether it is everyday selfishness or true egomania, does the most damage indirectly, by destroying the community that is essential to our work. That is the subject of another chapter. But as I suggested in the previous one, it is also obvious that money, status, and audience approval can lead us to choose our projects poorly, to capitalize on sensation and ignore depth. More audiences want to hear about cloning than about whether the human genome project was oversold. But dealing virtuously with this pressure is not a simple thing, not just a matter of saying, "I will resist this opportunity to satisfy my desire for applause and money." The deeper issues are those of self-deception and of discernment. An audience's attraction to the dramatic, after all, is a simple fact of human nature. Often it is possible to tap this attraction, and direct the energy in it toward deeper questions. Without the attraction, there might be no possibility of getting their attention at all. Finding the balance demands many kinds of moral attention: to the difference between what matters and what doesn't, to the limits on one's ability to redirect attention, and to the ways in which one particularly and personally is likely to be seduced. Approval, power, status, and money influence different people differently.

What makes this question interesting is its complexity. These external goods are often not purely external; they typically do not work in the simple way a bribe does, where a reward is available for doing the wrong thing. Accepting, and even seeking, these rewards can in fact contribute to doing the right thing. I gave the example above of seeking and using the attention of audiences in order to accomplish worthwhile things. The man who sought more office space may well have argued that in his institution, status is necessary to power—say, to the ability to strengthen the presence of ethics in the medical school curriculum. Status depends on perception, which in turn is affected by the kind and amount of office space you have.

Money creates other situations in which self-seeking intertwines with a desire to be effective. In an earlier chapter I described my surprise at the stipends that are customary in the field. I remain struck by the contrast between bioethics and philosophy in this regard, if still more struck by the sums major public figures charge. (A cabinet member, for example, said to have gotten $200,000 for a talk, may have received "only" $80,000.) But I have never found a clear way to think about these issues, and they are rarely discussed. Someone with whom I talked said there was a pecking order within her unit for paid consulting; I suspect the pecking order is implicit rather than explicit. So are decisions about whether and how much to charge. Obviously one must consider how much time will be required for preparation and travel. The resources of the inviting party also matter: some (like philosophy departments!) have little or no funds. But should one ask for a handsome sum from, say, a group of surgeons, knowing they could and would pay it? Is it a matter of what the market will bear? Is it a matter of respect: people only value what they pay for? The people with whom I talked differed widely on these questions. One, paid (and paid very well) for serving on a corporate IRB, believes she works more conscientiously because of it. Another described how he dealt with speaking invitations: "I tell them that my ordinary fee is $5,000. When they gasp, I say that I make exceptions for worthy causes or important audiences. . . . We talk about it, and reach an agreement. In fact, I never get anything like $5,000. But this way of doing things makes everyone happy: they think they've gotten a bargain, and they'll beat the bushes to be sure there's a good audience. I'm happy about the money, but also about the audience. Why speak if no one comes?" For another person with whom I spoke, money was unimportant. Like the person just quoted, she was a tenured academic. "I'm easy. For ethics I'd do anything. My stipends are low, $250 plus expenses if I'm visiting a campus out of town. I donate the royalties from my textbook to a scholarship fund for needy students, but most people don't know that."

Another believed that being paid for your work is essential if clients are to understand that "this is a professional matter." His fee was typically about $150, a figure he knew was low compared with what medical professionals charge, but high enough, he felt, to ensure being taken seriously. His fee varies according to the resources of the client, and he does some work for nothing (pro bono, for the public good), as he believes everyone should. On the whole, when the activity is ethics consultation, he prefers an annual contract, generally about $3,000, because it gives the institution more incentive to make use of his services. He would be interested in becoming a full-time ethicist within some institution for the same kinds of reasons: that he would be in some sense an outsider (as a nonpractitioner), and so able to raise the awkward questions; but he would also be an insider, and able to make more difference. He might be appointed to positions of real influence. He would ask for a somewhat higher salary than he gets now as an academic, because hospitals generally pay more, and he would need to be seen as a professional. But he also worries about the built-in conflicts of interest, the need to please the boss, to deal with internal politics.

The three ethicists quoted above differ in their attitudes about money, one of bioethics' external goods. Although I have problems with the deception used by the first, my concern here is whether some ways of dealing with fees threaten the internal goals and excellences of bioethics. Clearly our decisions about compensation can influence our choice of topic (the point I made in the last chapter) and of audience. We speak, teach, write, and so on in order to make a difference, and in order to learn from those who respond to what we say. Ability to pay is no measure of the value of any of this. Each of the people quoted earlier has found a way to be available to audiences with little or no resources.

There is a third way in which stipends raise moral questions. The money we receive rarely comes from our listeners' or clients' pockets. It ordinarily comes from their institutions, either directly (when the hospital, say, sponsors a speaker) or indirectly (when, say, a professional association pays us, its money coming from member dues or conference registration, these in turn reimbursed by the attendees' home institutions). The moral dimensions of this are twofold. The first is a new version of a point I made earlier: some people will be shut out of the audience. For philosophers, historians, and others in ordinary academic settings, the registration fees of professional and professional-ethics conferences are prohibitive. Whereas the American Philosophical Association's annual meeting has a registration fee of $25, the American Society for Bioethics and Humanities charges about ten times that. A similar ratio holds between statewide meetings of the Michigan Nurses' Association and the Michigan State Medical Society. At one bioethics meeting of the latter, nurses were particularly welcomed—but the meeting charge was high, and it was held at the Grand Hotel on Mackinac Island, perhaps Michigan's most expensive site. Very few nurses attended.

This conference no longer takes place on Mackinac Island. The fact that it did for a few years is partly explained by the fact that highly paid professionals—doctors or others—expect to attend conferences that cost a lot and are in glamorous locations. That's understandable, and possibly defensible. But when these meetings are paid for, ultimately, from health-care budgets, and effectively exclude some audiences, the custom poses significant questions.

Speaker fees, of course, are only one small part of what makes bioethics conferences expensive. I want simply to note that money we earn for speaking can have moral strings attached. The size of a stipend can skew our choice of question and of audience, certainly affects the way we spend our time, and can be part of the problematic practice of holding expensive medical conferences.

Instrumental Goods: The Quest for Funding

For most of us, however, the question of speaker fees is quite minor. More serious issues arise when our work depends on getting money from funding agencies. Some units are entirely dependent on what is called "soft money": money from funding agencies or sponsoring institutions for particular projects, money that will no longer be available when the projects end. The contrast term is "hard money," as in a salary line within a university, a hospital, or a professional organization. The distinction can be deceiving; a social scientist within an academic center reported that "Tenure doesn't mean much here. There's so much pressure to make money from patients or grants." Tenure kept him from being fired, but didn't guarantee him much in the way of salary. He had to earn it through successful grant applications. The situation is even worse for people and units completely dependent on soft money. In these situations people must constantly reapply for funds, often for new projects, knowing that many or most of their applications will be rejected. In many fields having one-quarter of your grant applications succeed counts as an excellent record.

The situation within a medical center, as I described in the last chapter, is especially complex. The great and understandable pressures to bring in grant money obviously affect our choice of project. In the last chapter I described responses when large grants became available through the Human Genome Project. I've heard essentially the same thing in other contexts, and I am not always cynical about it. One person described the largest grant her small center had ever gotten: "Over a million dollars. We were the only ones to apply and we got a very high score from the evaluators." (Without the high evaluation, they would not have been funded; allocated money is not always spent.) "We think others missed the opportunity because the RFP [Request for Proposal] included ethics as one possible thing in a long list of otherwise purely biomedical stuff." The issue was closely related to topics she and her colleagues had worked on for years, which is why they were reading the RFP in the first place. Nevertheless, they worried about whether this large sum of money would distort the unit's priorities. Could they manage such a large project and still give other things due attention? On the other hand, they saw immediately that their status within their home institution was suddenly much higher. "We're really visible now. People pay attention."

Things do not always work so well. A social scientist spoke quite soberly about the effects of grant seeking. "It's not just [the ethics money in the genome project] that drives research activity. Wherever the money is found, people will be working on trivial projects that they recognize as meaningless." She shared some morality tales. A scientist she knew had been doing work he found important, as (he thought) did everyone around him. "Then the funding dried up. 'Guess you'll have to work on urban rats, or fold up shop,' was the response. No one had really cared." About his work, that is; they clearly cared about the money he brought in. A nurse described a similar situation after a flow of money ended, in her case because projects were completed rather than because fashions in funding changed. She knew of a number of grant-supported efforts to help doctors and nurses learn to work together. None, she believed, had really worked, although many of the projects reported success. "Once you get beneath the fancy words of the reports, you see that nothing much really happened." Furthermore, the efforts themselves stopped as soon as funding money went elsewhere. These funding agencies wanted to promote significant, permanent change. But what they promoted instead, because it's the only thing they could actually reward, was simply the effort to get money.

All of this amounts to a constant pressure to compromise. One social scientist described someone who managed to resist, a researcher who believed that only he could explain an experimental protocol to prospective research subjects in a way that guaranteed that they really understood what they were agreeing to. But "after the research was done, he would call the subjects back in and say, 'Here's how the experiment turned out.' Almost invariably they said, 'Experiment?' They just hadn't gotten it. *And he gave up all such research.*" The emphasis is hers; she had never known of anyone else doing anything like this. (In contrast, she mentioned research into the effectiveness of ethics consultations: "You never find that people would be willing to stop doing the consults if that's the way the evidence came down.")

The activity of applying for grants poses a number of moral challenges. The first is to try to find a way to deal with what matters, either by working without funding (not possible for some) or by diligently searching out the RFP or the small local agency that matches your own work. One center adopted a policy of agreeing to do paid consultations only when they believed they could learn from doing it, and in general do not accept such jobs.

Next there are challenges to honesty. Suppose you believe your work is important, but not for the reasons the funding agency does. One social scientist, explaining a long-term project involving genetics, confided that "we will probably never find" the results for which the team was funded. "But don't tell that to the funding agency." I couldn't have told them, of course, but more to the point, she and her team won't. The grant application may well have been written deceptively, and this does not seem to be uncommon. A certain amount of deception or at least fabrication is also built into the reporting of grant-supported activity. I once heard a foundation official laugh fondly at the naiveté of an applicant who returned part of the money she had gotten, when she was able to finish her project early. Generally one finds a way to use all the money allotted. Another rule of thumb I have heard is that a project needs to be half done before one can write a decent proposal to begin it; once you get money to "begin," you should spend half that funded time preparing and submitting grant applications for further projects.

One of the people with whom I talked understood the pressure to get grants for the overhead money and the status it brings a university, but said, simply, "I won't do it." He had gotten trapped once; administrators like to use his name on grant proposals because he's been successful and is well known by one of the major agencies. For a while he let them do that, until he found himself committed (by the terms of an accepted grant application) to do something he couldn't: travel overseas during a month when he had prior obligations at home. I myself once wrote a grant promising my time during a summer when I already knew I would be teaching in London. I suppose the tactic could be called bait and switch; fortunately in my case, the bait was not sufficiently attractive and the application was rejected.

There are also incentives to engage in manipulation. In grant workshops one is told to develop a relationship with a staff officer, which can be useful and straightforward. But bad faith is another possibility, and it's hard to separate the two: "I want time with you, not because you know much about what I'm proposing, but because you know what pleases referees. And because you have power." Workshops will also suggest that one "schmooze" with the university's "development officer"—someone whose job is to persuade donors to give or bequeath money to the university. "Convince them they want to support you. They'll ask, 'What are you doing that's interesting, that we can sell?'"

On the other hand, there are also bad reasons for *not* applying for grants. Dealing with funding agencies can be time consuming and miserable. Applications can be hundreds of pages long, budgets alone taking ten or twelve pages, half in narrative form ("We will spend . . . because of . . ."), the other half figures and computations, including such things as postage over the next three years (300 mailings @ $.34) and fringe benefits for graduate students doing a one-quarter-time assistantship during the summer, prorated for inflation; and so on and so on. Dealing with agency personnel can be another irritant. Typically funding agencies have a professional staff assigned to help applicants put together good proposals. These staff members also serve as gatekeepers, giving a little feedback to everyone who asks (possibly months after they ask), but choosing a few promising ones to groom for the panel of referees. So, after perhaps six weeks of full-time effort putting the proposal together, sending in a draft for feedback many weeks before the deadline, applicants wait. There may be a phone call five days before the submission is due (in twelve copies, with a financial cover sheet signed by five senior officials scattered widely around the university). The phone call suggests extensive revisions, and of course you will do them, for the phone call is evidence that the staff officer will be pushing your proposal. Or you may open an envelope unsuspectingly a month or two later, to find a form letter stating, "We had many fine proposals. . . . [U]nfortunately we were not able to fund yours."

The process is a humanly messy mix of bureaucracy and personality. One scientist spoke with anger about it. She had served on the panels of experts who evaluate applications. "I saw a proposal from an ordinary institution turned down because they hadn't specified their lab capacity. Another proposal from a more prestigious institution failed to do the exact same thing, but it didn't matter. The other judges just waved the problem away. 'Science is in the air there.' Applicants don't know this happens and they have absolutely no recourse." Another researcher estimated that she spent fully half her time writing grant proposals and evaluating the proposals of others—not the life in science she had once imagined.

In the face of this confusion, the moral challenge is to develop the courage, the humility, and the compassion to keep applying when that is the right thing to do. The temptations are self-righteous sulking and "principled" withdrawal from the whole activity, even when truly important work depends on it. On the other hand, principled withdrawal *is* sometimes the right choice. A social scientist remarked, "It's a fantasy that this is the only way to lead a research life. . . . [F]or one thing, the salaries [supported by certain kinds of grants] are enormous, much higher than people really need."

Deciding whether to seek grant money, then, is morally complicated. There are various lines to draw, some of them fine. Does drawing these lines matter more in bioethics than in other fields? I believe it does. One of my interviewees, speaking of all our activities and not just grant seeking, said, "This role calls for circumspection. Bad role modeling by an ethicist can do real damage." Like others with whom I've spoken, he knows the arguments that say ethicists need not be particularly ethical people, but he is not convinced.

Those who think that what we do is sheerly moral reasoning, the crafting of excellent argument, are probably most comfortable in claiming that our character is irrelevant to our work. Those who describe our work more broadly, as I do, will be more inclined to think that the way we live matters to our work. Even on the moral-reasoning model, however, it matters that we at least think seriously about the discriminations that everyday life requires, because the best ethical thinking is grounded in the complexities of the world.

So far, I have focused on the choices we make as individuals or as institutional units. It is equally important that we take part in scrutinizing the whole enterprise of getting and giving grants, to enter into the arena of science policy:[2] What should national priorities be, in terms of goals to be sought and programs to be supported? Should only a handful of universities, for instance, be expected to do serious research? Should there be more differentiation, so that on any one topic fewer campuses are competing for funding? What kinds of communities do current practices support, within and across institutions? What questions get overlooked, given the way research is funded? (For example, an epidemiologist I know argues that funding in his field supports attention to risk factors in individuals, but not to social patterns, as causation and as sources of intervention.)[3] Because biomedical research is profoundly affected by the way these questions are answered, bioethicists must be involved in the discussion.

Internal Goods: The Need to Be Heard

Still deeper problems in bioethics are neither external nor instrumental goods, but what I will call internal goods, those which by the very nature of our work we must seek. One is the need to be heard: by clinicians, administrators, policy makers, and others whom we hope to serve. Usually this implies a need to please; otherwise, we will no longer be invited in. The people with whom I talked often mentioned this danger and told stories to substantiate it: "In the 1980s I was an external consultant, and advised that artificial food and hydration [tube feeding] could be discontinued. I was unpopular." "The Hospital Ethics Committee suffered terribly; it disagreed with a doctor, and the doctor won. The committee went down the tubes." "He was very unpopular; he speaks out." "I was silenced. I could never be allowed in the NICU again."

Doctors irritated by an ethics report have forbidden anyone working under them to call for an ethics consult. Administrators threatened by an unsolicited ethics committee report can block the whole initiative. In a medical school the ethics curriculum might be cut in half. On a policy-making committee, those whose comments are unwelcome may not be invited back. Such committees themselves, if their recommendations are poorly received, can be dissolved. In all these arenas, the desire to make a difference in the long run can make us hold our tongues in the short run. To some extent this is wisdom, but at some point it becomes cowardice.

We need to be heard not just locally but nationally. The process by which editors accept and reject manuscripts is also morally intricate, involving among other things the ownership of data, which may be a matter of institutional or grant-agency policy. (A recent case in the news involved a decision not to publish results showing that antibiotics made no difference for children's middle-ear infections, when the children were examined eight weeks after the illness. The researcher who "owned" the data, because he was officially in charge of the project, chose to publish only the results found four weeks after the illness, when those who had been given antibiotics did slightly better. The study was funded in part by a drug company.)[4] Someone with whom I spoke described an apparently much worse situation within bioethics. His team had looked at research done within nursing homes and found many ethical lapses. The paper reporting these findings was rejected for publication (I believe by a medical journal). The reviewers thought publishing it would make doing research in nursing homes still more cumbersome, and would decrease the amount of funding available for it. What does one do with important but unpublishable results? How much effort should go into making them known? How much danger to oneself need one accept? The researcher into middle-ear infections essentially sacrificed his career in the effort to be heard. I do not know what road the researcher into nursing homes finally took.

These conflicts, then, can be deep. They are often a matter of occupying an institutional role which, for good reasons, is limited, while at the same time trying to recognize and respond to a broader set of obligations, to occupy a professional role that religious discourse calls prophetic. On the one hand, ethics work is not police work. In hospitals, for instance, ethics committees enter particular cases only when invited, and offer an opinion rather than a decision; they do not want to be seen as an enforcement body of some kind, and their work—which so intimately involves helping others think, decide, and act—would be at risk if they were seen as prosecutors rather than as a service. Yet there's something paradoxical here as well. If in any sense an ethics committee should be the conscience of the hospital, this model won't allow it. A conscience is supposed to *raise* questions, not just help resolve them. There have been cases of flagrant ethical abuses within institutions where well-known ethicists made their home. Did they know? Should they have known? While the answer would be different in each case (and I have no opinion about any of them), the answer cannot be an automatic "Of course not."

There are other barriers to raising unwelcome questions besides fear of retaliatory exclusion. A bioethicist can simply understand his role in overly narrow terms. An example appeared in the *Journal of Clinical Ethics* a few years ago. Two adolescent girls, twins, had been recruited for research involving a lumbar puncture (LP). LPs are painful procedures carrying some risk, and were to be done for the sake of research, not in order to help the girls. (As is not uncommon, however, there were unspoken expectations that doctors outside the protocol might use the results in handling the girls' problems. This could not really be said, but was understood, and muddied the picture.) The girls did not seem to understand the situation fully, and were being badgered by their father into assent.[5] In her comment on the case, Laurie Zoloth criticizes the ethicists involved, who seemed "merely to describe the cascade of troubling circumstances without halting them, or at least calling out the questions of motives, relationships, and essential duties. . . . It is in part the role of the ethicist to . . . raise the difficult questions that stand just outside of the therapeutic encounter."[6]

Let's assume that Zoloth is right, and that these ethicists failed.

(As she notes, we are dependent on a written account; there must be other ways to tell the story.) The ethicists felt discomfort but remained quiet because of the way they understood their role. As the written account puts it: "The ethicists agreed that the twins had understood the particulars of the protocol and had assented to participate. The ethicists were uncomfortable with the family dynamics and the atmosphere in which the assent had been obtained, but, as the conditions imposed by the IRB had been met, and the informed consent process had begun before the ethicists were called, it did not seem appropriate to stop the process."[7]

The strain they felt will be familiar to many in bioethics. One might, for instance, tell a physician about hospital policy on unilateral DNRs, because that is the question he has asked, but believe that he will misuse the information. Should one speak up, or be proactive in some other way?

Whole ethics committees can develop questionable habits of reticence. Someone described being pressured by his colleagues: "I wanted to put into a consult report, 'The cardiology service needs to look into this.' Two senior members stopped me: 'Don't say that. It could cause legal trouble.' But then how does a hospital reform itself?" One way of working toward the integrity and discernment such situations require is the use of an electronic discussion list, like the one currently hosted by the Medical College of Wisconsin. In that forum people can bring up the particular practical problems they're facing, and hear how many others would or have dealt with them. It helps people not only with reflection, but also, in Whitbeck's term, with design: In practical terms, what will work in such situations?

So far, I have described the things that keep us silent. The complement of this intimidation is co-optation: being free to say anything, because speaking up will accomplish nothing, or rather, the wrong thing. I heard several stories to this effect. One report came from an IRB member: "Talking about a particular proposal, I asked: 'Shouldn't people be told about the ultimate financial implications?' Everyone else said, 'If we did that, no one would sign up [as a research subject].' So my objection was overridden. This kind of thing kept happening. Eventually I decided I was the nominal face of ethics, my being there served to convince the others of their own good conscience—and that was all."

Another story concerned what might be called organizational ethics. "My institution was planning a big innovative facility where most of the care would be provided by family members, in apartment-like settings. Great cost-cutter for the institution, because so few nurses would be needed. I pointed out that this favors people with family members who could take time off from their jobs—which is the people who are already doing pretty well in the world. The only result was that the chief planner, in media interviews, could bring up those issues and say, 'We're sensitive to these issues.' It established his ethical sensitivity and absolved him of the need to actually do anything. And I had given him the words to do so."

A third example was resource allocation: "I'm 'the ethicist' on a commission doling out funds on AIDS: they make hard resource-allocation decisions, like whether to allocate money for measuring viral load, and if so how often; should we fund treatment or measurement; and so on. They want me to 'bless' their decisions. There really isn't any particularly right or wrong answer for a lot of these things." "What do you say?" I asked. "Nothing intelligent; some things about procedure, whatever."

In the first case above, the ethicist resigned. Again, these are not easy calls. It's possible that the IRB without him will do still worse things. But personal integrity might be more important than the slight chance of making a difference. In each of these situations, the person talking expressed a discomfort ranging from uneasiness to disgust.

Finally, there are cases where one is heard with all too much effect. Our purpose, roughly, is to help others see and reflect upon the moral dimensions of a situation, and to act accordingly. Our role is not to decide what should be done. In ethics consultation the line between the two can be hard to maintain. Consultations can sometimes (often?) be seen as a request by doctors for permission to allow a death to happen. Some ethicists take this as a good thing, a move from the days when doctors made such decisions alone. Ethics consultations reduce the chances of an abuse of power, and lessen the agony conscientious doctors endure. But not everyone finds this a good thing. "The docs can sleep better at night [after an ethics consultation affirms their decision]. But this makes me uncomfortable. It's not my job to give people permission. I really emphasize that I can't do that; the most I can do is reflect their own thinking back to them, and situate it in the national consensus on bioethics. We need to raise consciousness without being judgmental."

This ethicist feels he can continue doing ethics consultations, even though what he offers is different from what some who consult him want. Others find themselves in more deeply compromising positions. Carl Elliott writes of being asked during consultations simply to make decisions, a function he has painfully learned to understand as the way some institutions function, the reason they want ethicists in the first place, but the role makes him extremely uncomfortable.[8]

Elliott's experiences, like those of the others quoted here, demonstrate how accomplishing an apparent or partial good can interfere with accomplishing our real goals. We want to be heard in such a way that listeners think more clearly about the values at stake, choose (not just accept, but choose) a morally defensible course of action, and pursue it effectively. For this to happen, we must be both agreeable and challenging, while remaining alert to the possibility that our words are serving the wrong purposes—perhaps providing a useful illusion that certain values are being honored, or a welcome excuse for those who reject the burden of decision.

A life in bioethics, then, requires integrity and discernment as we sort out the real goods from the apparent, distinguish temptations to self-aggrandizement from proper self-regard, and find ways to protect ourselves and our projects that do not vitiate them. These moral challenges confront us not just as individuals but also as communities. For this reason I turn next to the question of what makes a community virtuous.

9 Virtuous Communities

We all work within some kind of community, and the nature of those communities matters. It matters, first and most simply, because human beings should treat one another decently. Virtuous communities are also required for the sake of our work, which flourishes in some environments and withers in others. An example comes from Chester Burns, a pioneer in the field, who reports that it originally took him three years to "get up the guts" to give a speech saying the Hippocratic Oath was insufficient for medical ethics. The president of the institution, the man who had hired him, didn't speak to him for three months afterward. "I thought I was going to be fired. I really did."[1] *But he took courage from his colleagues.*

We can supply one another not only with courage but also with many kinds of emotional, moral, and intellectual support. Earlier I mentioned a nurse-manager who feels "a real lightening of spirits" among his staff after working with them on mutual respect; "it frees up energy to give to the patients." Conversely, when support and respect are absent, it can be impossible to work well. One physician described a medical unit she was leaving. It was far from her first experience with stress, but she had never seen this degree of chronic anger. She remarked, "If in a fundamental way you can't take care of yourself, you will take it out on the patient." Like healthcare professionals, bioethicists can fail those they should serve, and can finally be driven to leave the job or even the field. Several people with whom I spoke described their decisions to move as "mediated by toxicity."

Research confirms what these anecdotes, and common sense, suggest. When Cary Cherniss studied occupational burnout, he found that supportive organizations become self-perpetuating, as good workers remain and provide an environment that encourages others to remain. He counts as supportive the organizations that allow professionals a great deal of autonomy, recognize and appreciate what they do, allow flexible schedules so family demands can be met, and provide opportunities to learn.[2] Of these, most people in bioethics have the first, a substantial amount of autonomy. The presence of the other factors varies. Cherniss also found that when the workplace has excessive red tape and too much internal politics—a problem in many hospitals and universities—people tend to leave. Finally, he believed that "moral communities," in which people share a moral vision and commitment, provide the best environment for keeping people engaged in their work.[3]

Along the same lines, and in an analysis that in many ways parallels my own, developmental psychologists Anne Colby and William Damon investigated the ways in which moral understanding and commitment grow deeper. They studied a group of people whom they called "moral exemplars."[4] My interest is not so much in these individuals—for I do not think bioethicists either are or need to be moral exemplars—as in the relationship between them and the communities with whom they worked, and the kind of growth the relationship made possible. I also take seriously the authors' claim that there are commonalities between the everyday moral lives we all lead and the "enlarged commitments" of the people they studied. Where "popular culture tends to dramatize moral heroism as if it were a magical fantasy," these researchers believed that everyone is capable of growth, and that understanding the development of preeminent virtue should be of use to everyone.[5]

Colby and Damon focused on the way in which people become fully committed to an ideal while growing within that commitment. Contrary to popular myth and stereotype, these exemplars "did *not* . . . endlessly reflect on what is right or wrong . . . constantly struggle with temptation, fear, and doubt . . . lead grim, joyless or dreary lives . . . [or] fight many of their battles in splendid isolation. . . . [They were not] harsh and unforgiving with themselves and their followers; [nor did they] provide their followers with a definitive, fully formed vision. *The truth . . . places moral exemplars far closer to the center of a collaborative support group.* It is a picture of striking joy, great certainty, and unremitting faith; one that results in both high standards for the self and charity toward others"[6] [emphasis added].

One might wonder whether this description fits with the kind of moral development I presented earlier, the growth that I hope we will foster in others and in ourselves. Is such great certainty consistent with moral reflection? Can people without doubts engage in moral reasoning? The answer is yes, once one distinguishes certainty about ultimate goals (often very general: help the poor, educate children, heal the sick) from the exemplars' constant questioning about how to pursue the goals. Note that the exemplars did not offer their followers "a definitive, full-formed vision." In fact, the authors argue, it was the very generality of the exemplars' commitment that kept them open to serious specific questions: not only about what sort of strategies would be effective, but about which ones fit their fundamental moral commitments. The goals themselves developed and changed, as the exemplars rethought, for instance, what really counts as "help." Similarly, we in bioethics share a general moral commitment: to improve health care, health policy, and the public understanding of both. We differ vigorously on how to specify and how to pursue those goals, and our differences are useful.

Colby and Damon's exemplars, then, were people whose certainty about fundamentals promoted openness and critical self-reflection throughout their lives, well into old age. The authors called this a "developmental paradox" and set out to understand it more fully, and to study it developmentally: to study, that is, not only a set of characteristics, but also the ways in which they were acquired.

What the researchers found was a second paradox. The exemplars, independent and self-directed people, constantly drew from the communities they helped form; they drew not only strength but also perspective. Their relationship to the communities that sustained them was one of mutual challenge as well as support (which is why vigorous disagreement within bioethics is helpful). Examining the relationships of exemplar to community further, Colby and Damon discovered the following:

The exemplars had collaborative relationships with—worked together with—people whose points of view partly overlapped, partly diverged.
The relationships continued over extended periods of time.
During this time "rich and frequent communications" took place about underlying values.
Eventually exemplar and community adopted some of one another's goals and strategies, often modified to fit within differing fundamental perspectives.[7]

Community, then, can supply more than the pleasures of companionship and the strength of numbers. It can also promote moral growth, by offering a shared but always enlarging perspective. The idea is reminiscent of Aristotle's highest stage of friendship.[8] Obviously, too, although not the focus of Colby and Damon's research, people working together usually accomplish more than people working alone. Those working together can draw upon shared resources of competence, time, and energy.

Not surprisingly, some of the communities formed within bioethics further our work, and others do not.

Degrees of Community

Communities within bioethics take different forms: day-to-day collegiality within an institutional unit; commitment to a shared project among task force or committee members; the shared if diluted allegiance within a large organization; collegiality within professions and disciplines. These levels can compete with one another, as when identification with a national professional community leads someone to ignore local needs and possibilities.[9] On the other hand, community of any kind can make up for deficiencies in others. As one man said, "I work alone for the most part. I'm the only one in the institution with this job description, and even in town there's no real community of ethicists. But I do have a very significant conversational partner from [another department]. She keeps me honest, moves me beyond the perspective of the privileged. My wife and I socialize with her and her husband; we all have social justice interests."

This man has no official community *at* work. But *in* his work he does, and that community fits Colby and Damon's model: its members challenge one another to grow.

More structured communities can do that, too. One person described his own unit as "pretty collegial. Usually we take on projects as a whole team, practically everyone works on them. Once we had a new director who suggested [a new approach:] 'Everyone eats his own kill.' We absolutely rejected that." The new director meant, essentially: make your own connections, create your own projects, get your own grants. All the profit from doing so—money, equipment, connections, knowledge, publications—will belong to you. Perhaps he had a simplistic capitalist model in mind, one that values the entrepreneur and misses everything else. He seems to have been blind to what recent theorists call "social capital": the ways in which a group of people makes better use of resources if the group is characterized by mutual trust and sharing.[10] The startling metaphor "eat your own kill" reminded me of hunting and gathering societies, but what I remembered was a contrast. In such societies there can be elaborate rules for the distribution of every large animal killed: the heart goes to the hunter, perhaps, the haunches to his mother-in-law, and so on. Whether "kill" means meat or any other kind of resource, it can be a means of growth and strength for the whole group.

Not everyone understands this possibility. One person described a unit lacking this kind of sharing: "We're basically independent contractors. Everyone finds and works on their own project, for the most part. We share a secretary and a photocopier, but that's about it." Some members of this center found community outside it; others worked in desperate isolation. A young woman I'll call Jamie worked in a similar situation, and suffered significantly from it. As her name suggests, she was a generation and more younger than me; I came to know her when I was beginning this book. I learned a lot from her experience, and hope I was some help in the struggle she had to wage. Her experience contrasted sharply with my own as a junior faculty member two decades earlier. As a newcomer she needed to be introduced, integrated, and informed in all the usual formal and informal ways. Her professional development was severely hampered for years because no one thought to do these things for her. At my prompting she began to speak up and ask for help, but that, too, fell on deaf ears. She thought there were a couple of unspoken assumptions from her colleagues: "We did it [made connections, developed projects] on our own. If you're any good, you can do it, too." In the end, she did (and got some help from unexpected sources). But during the years it took that to happen, it wasn't only she who suffered. The unit as a whole was less than it could have been.

Someone described this kind of structure as "parallel play." The term is a half-humorous, half-insulting comparison with the way toddlers interact. The children may be playing with the same kind of toy, but they ignore one another; in particular, they do not share. Children grow beyond this stage. Obviously some kinds of work, even when done by adults, demand this kind of detachment. Academic work in the humanities, where most serious effort takes place alone at the computer or in the library (and most of what is written has a single author), is one example. But this apparent isolation often conceals real and important communication. Even someone working on Fermat's last conjecture would find it helpful to have someone down the hall interested in the same thing, with whom to share an intellectual knot, an article, or a search for a forgotten reference. Contemporary workplace designers know this, and try to build in the chance for unplanned encounters and casual interaction.

Some kind and degree of local community, then, is very helpful and even necessary. But it doesn't always exist; or it can exist but be destructive. One woman who found herself for the first time in a predominantly male unit cried, "I can't stand the atmosphere around here. I feel as if I'm suffocating." Her colleagues did work together to some extent, share interests and projects. But that was all; what caused her such distress was the lack of emotional community. "No one talks about their children! No one talks about anything except work—if that." She had just learned that the wife of one of her colleagues was terminally ill; although the diagnosis was not recent, this was the first she had heard of it, and she thought she was the only one in the unit who had. Her colleague was suffering alone, in the only way he knew how.

I asked someone who had worked in both male- and female-dominated situations whether she thought gender made a difference. She said, "With women, work relationships tend to be life-long friendships, involving people's kids, celebrating life events. And there's absolute candor about why we can't do something. It's understood that people get interrupted if their kids call. They take the call. We're always on a first-name basis. There's food at meetings, making them into social events; somebody just brings it." This is not to say that all men, or all women, are alike. The woman who felt suffocated later found a soul mate in a male colleague; another woman whose male colleague canceled a meeting to take care of his child commented angrily, "I would never have done that. I *could* never have done that." The clear implication was that he should not have, either.

Nevertheless the question is interesting: are emotional and personal connections at work necessary, or helpful, for doing good bioethics work? They can certainly be dangerous. Friendship can become hostility; the fuel for it, in terms of knowledge and emotional power, is immediately at hand. In addition, closeness between some can exclude others, and will inevitably change the dynamics of the group in some way. On the other hand, professional and personal relationships could not be cleanly separated even if that were desirable. People who work together learn personal things, about temperament, work habits, sensitivities. Professional conferences are important for the chance to talk with friends from elsewhere, talk that moves seamlessly from "I've gotten married!" through "I'm up for tenure" to "I'm working on this new project."

Our current construction of gender, I would argue, provides a compelling argument from justice for mingling the personal and the professional. The women's movement has brought about some fundamental changes in the workforce, but other basic social arrangements remain unaltered, the most significant of which is child care. It is still the case that men usually take much less responsibility for it than women do. In addition, professional lives demand more, rather than less, time than they did decades ago.[11] The result can be crushing to women with children. Many of them, like women all over the world, work "double days," one on the job, one at home. They either agonize with the burden or cut their professional commitments for a while. That would be fine if fathers did the same thing, and of course a few do; but only a few. The result is that women with children tend to fall behind men in their work lives, and it can be impossible to catch up. In fairness to these women, and to a field that needs all the perspectives and energy possible, the bioethics workplace should acknowledge that children exist and need to be cared for.

Units can change, toward and away from real community. One person traced a deterioration in this way: "A few years ago we had a communal coffeepot. Now everyone has their own, in their offices. Maybe the change started when there were two pots, one for decaf, one for regular." Coffee nudged my own center in the other direction, toward more collegiality. When a shop opened nearby, we found ourselves dashing out together to pick up lattes, and talking about our work along the way. Coffee is hardly the most important reason units change, but the sharing of food and drink almost always helps.

One of the most important ways in which communities contribute to moral growth is through regular, challenging conversation. What I've said so far addresses the fact that in some places people hardly talk at all, and so this possibility is unrealized. The much more troubling situations are those in which people talk, but not freely. I saw that in two sorts of situations: where religion was involved, and where job security did not exist. In one Catholic institution someone who firmly identified herself as Catholic said, "Being an ethicist here is tricky. The climate has chilled over the past few decades. . . . In the end I don't think my choices are different, but the stakes are high. Signing a petition about gay rights on campus could be a career-ending move. . . . [On the other hand] I can't speak the party lines." This woman understood that for the church the additional constraints are expressions of principle, but she felt not only at risk but in a sense bereaved. "Stifling dissent is no way to revitalize an institution."

The wrong kind of secularism can also be stifling. Bioethics seems to me marvelously open to religious perspectives, greeting them with respect and interest, but my marvel may be largely a function of the contrast with philosophy.[12] Others do not share my perspective. One, for instance, reported a noticeable cooling from colleagues when he let it be known that he had worked with his presbytery on issues in bioethics. For my part I hesitated to identify the institution above as Catholic (the tradition in which I grew up) because I know how deep and unfair anti-Catholicism can be.

Another situation in which I saw dissent muzzled was among salaried, nonunionized, hospital employees: nurse managers, chaplains, social workers, and so on. Of course this differs from institution to institution, but seeing the fear that permeates some work lives sobered me. A courageous ethics committee is never more important than in providing legitimacy for disagreement. In a previous chapter I used Margaret Urban Walker's phrase "keeping moral space open," and focused (as she does) on "moral": these issues can be lost within a forest of technical and pragmatic ones. Here I want to underline the other concept, openness. In some hospitals *open* space will not exist unless an ethics committee creates and maintains it.

So far I have talked about qualities of communities as a whole, about cultures in which people talk or do not, speak freely or do not. Even the most open community, however, can have individual members who are marginalized—not part of the main story.[13] Everyone suffers when this is the case, but the person pushed to the side suffers most of all.

On the Margins: Junior Faculty and Others

Many things can make one marginal: being a newcomer, being young, belonging to an ethnic minority, being the only woman or the only man, being a citizen of another country. Some of these statuses bring with them special kinds of insecurity, particularly the job insecurity of the young, the ingrained gender assumptions faced by women, and the habitual exclusions endured by people of color. An African American said to me, "I attended [bioethics] meetings year after year. Often I was the only person of color. Only a handful of people ever spoke to me. Then finally people started to think diversity was a good thing."

When we invite the new, the young, the minority, the foreign national, to join us, we need to consider whether they need special help at first, as Jamie did. Built into these statuses is not knowing how things work, an ignorance which cannot be remedied by reading, only by working and living with those who do know. There can be a special and terrible irony built into what are called "affirmative action hires," the hiring of someone in order to add diversity of gender, race, or nationality to the unit. The irony is that those hired *because they are different* may well differ in more than skin color or gender. Almost by definition they will not have the same set of tools insiders have; their experiences, networks, and competencies will differ. If we want to gain from the diversity they provide (and not just be able to point to our own virtue or check off a box on a score sheet), we will have to share our own competencies, probably for a number of years. Introduce people to the (usually complex) personalities and dynamics of the larger institution, get them included in ongoing projects, think (and ask them) about what experiences they need. Don't assign them anomalous or highly controversial projects just because "someone has to do it" and their time looks free.

Valuing difference, we may be blind to the fact that those who are different will not flourish, may not even survive, unless to some extent they become assimilated, a state it is almost impossible to attain on one's own. Someone who was different in several of the ways I mentioned earlier reported of her first months on the job, "I came in early, worked all day—no one said good morning, no one said good night. They don't seem to talk to one another, let alone to me. The director is never here. We never have meetings. Once a couple of them were talking about a project, I think, in the hall, and I sort of wandered up and tried to join, but they looked at me as if I were intruding. So I left." In addition to the traits that marked her as different, she had had the bad luck to join a workplace where "parallel play" and "independent contracting" were dominant, although the same exclusion can result wherever one group, even one pair, dominates a center's life. I saw the same sort of desperate isolation in a case of a "spousal hire": someone who had been given a job in an ethics center because his wife had been hired in another part of the medical school. Such people can be abandoned as surely as the young can be, resentment perhaps compounding the problem.

One night an old friend (not a bioethicist) and I watched a video of the movie *Apollo 13*, a true story in which heroic teamwork saved the lives of astronauts stranded in space. My friend had recently retired from military service, and when the film was over she said, in tears, "That's male bonding. And I could never, ever, be part of it." My friend, whom I'll call Lynne, started talking about her life in the military. She had found little of the crude sexism that makes its way to the courts and the headlines. (In fact, she found her military colleagues considerably more civil than civilians.) Instead, she was talking about blindness, unintentional, unrecognized, and terrible. The story I remember best is simple, about a day when everyone except Lynne reported for duty in dress uniform; the others, all men, heard about the change in dress code because they shared sleeping quarters. She, the only woman, didn't get the word. In the military the uniform is a powerful symbol of unity, and what she wore that day dramatized her exclusion.

I was struck by the similarity between Lynne's stories and some of Jamie's. She had told me about finding out that a prominent bioethicist had stopped by her unit that morning. No one had told her; she had no chance to meet him. The visit was brief and informal, but not impromptu; her colleagues had known it would happen. During the visit one of them told the visitor about an interesting case, one thing led to another, and eventually the two (who had not met before) cowrote an article for a major journal. Jamie, at a stage in her career when she desperately needed publications, might have joined the conversation and the project—if she had even known the visitor was there.

In contrast, my own center, as part of its yearly evaluation, asks each person, "What do you want to accomplish? *How can the center help?*" [emphasis added]. That attitude is crucial, and particularly toward the young. Whenever I talked with Jamie, I remembered in frustrated contrast my own years as a junior faculty member at Old Dominion University (ODU), where so many people helped me.

(And I was particularly in need, having gotten none of this in graduate school; I suppose my version of Jamie's story happened at an earlier stage in my professional life.) At ODU, colleagues helped me learn about publishing, presenting at conferences, and about linkages with Women's Studies; the Women's Caucus presented a panel on the tenure and promotion process; a network of women kept one another informed about things like the opportunity to teach in the Honors Program; the dean suggested my office be moved because it was too isolated for a newcomer. In much of academia, such measures have now become institutionalized. Lynne had some of that kind of personal and institutional help, but not nearly enough. Eventually the alienation destroyed her health and her career. Jamie was at similar risk.

Nor was she unique. I talked with more than one young faculty member in unnecessary trouble. For those with university positions, tenure standards were usually unclear. The standards of academic medicine are wildly different from those of, say, history or literature. "I was told that I needed twenty peer-reviewed articles in refereed journals, and 'review articles' would not count. But Medline [the major data base for medical scholarship] calls every article without data a 'review' article." It followed that no article in her home discipline, where research was not quantitative, would be counted. Humanist scholars do not typically collect data. They might analyze the relationship between data and conclusion, or evaluate decisions about what data to collect, or look at the language in which it is reported; or they might do entirely different things, like trace the history of a hospital or analyze what justice requires in the matter of national health care. In Medline all these very different endeavors are collapsed into the single not highly respected category of "review articles." Who will value them? A physician with a doctorate in philosophy reported in frustration, "Yes, Harvard is publishing my book. And yes, I've had two articles in *JAMA* [the *Journal of the American Medical Association*]. But I probably won't get tenure. The philosophy department doesn't count my work as philosophy. The medical school doesn't count it as medicine." In a variant of this problem, the expectation within science departments that young faculty will be "first author" on a certain number of articles cannot be translated in any reasonable ways for work in the humanities. Being first author on ten articles is not the same as being sole author on ten, or five, or twenty. The two enterprises are not easily compared. But if there are no right arithmetical answers, there are a lot of wrong answers, which put young academics at severe and unjust risk.[14]

I did talk with one person with a joint appointment who found the disjuncture of standards useful. "The med school committee bowed to the Arts and Letters committee and—I think—it deferred to the medical school committee. Neither believed its own standards applied. So no one really scrutinized my case. And that was fine with me." This situation did not seem to be common.

Still another problem arises with the area called service. "Service" is one of the three facets on which one is evaluated for promotion and tenure, the other two being teaching and "research" (which includes many kinds of public professional output: analytic and creative writing, painting, and so on). Most young faculty are warned about the risk of spending too much time doing "service," which includes committee work, public speaking, sponsoring student clubs, and so on. In a sense such activities are required, but in fact they will matter almost not at all at tenure time. Putting too much time into service is a classic death trap for junior faculty, as one person's story illustrates: "A major accreditation report was due, and someone from our unit was expected to spearhead it. My director asked me; I agreed. It ended up taking most of two years' of my time. At the end I was told that since I had published so little, I was not even a candidate for tenure, and should look for another job." He had been hired with unclear expectations and an unclear tenure status. He did not survive.

The lack of clear standards and institutionalized support for junior faculty creates some grave injustices, which is the major reason for trying to change it. But the lack also hurts the work we do, which needs the voices and energy of the young. We should not be weeding people out in the random way that seems to be happening.

Underlying the lack of standards and support, and exacerbating it, is the fact that there are no standard models for bioethics centers, and so no basic structure into which someone transferring in can expect to fit. This is in contrast to someone transferring from, say, one history department to another, or one ICU to another. In these transfers, details and customs will differ significantly in the new unit, and one can badly underestimate the extent of the difference, but the fundamental structure is the same: so many patients, so many courses, and clear expectations about what it means to take care of them. Some centers approximate this by formalizing responsibilities so that a new person has a place to fit: Each person might be a liaison to certain clinical departments, for instance. But no structure will fit all institutional situations, which vary substantially. Until and unless some common structure develops, bioethics centers will need, instead, an acute awareness of the special problems faced by junior faculty and other newcomers. The basic moral issues here are not only respect for the new individuals, but also for the fact that our work will be done better when everyone is able to function fully.

Leaders and the Senior Establishment

I came to expect a certain almost comical description of the directors, leaders, and bosses under whom my interviewees worked. "He can't carry on a conversation without opening mail, sorting through messages, checking his answering machine, skimming his e-mail. . . . He checks his watch every thirty seconds. . . . He's always grading tests during meetings. . . . It's almost impossible to find time to meet with him." As soon as I mentioned one or two of these habits, the listener was likely to say, "That's Sam!" although I had probably never met Sam. I want to emphasize that probably no one commits all these sins, and many leaders commit none. Virtually all, however, seem to be overly busy and overcommitted, and that presents certain occupational hazards.

Almost by definition leaders in this field are high-energy, ambitious, goal directed, and politically skilled.[15] Beyond that, they vary broadly. Most of us could name without hesitation ten or fifteen people whose courage, energy, insight, and dedication have been indispensable, and within a few more minutes the list would go on and on. These are not saints, but human beings who have used their considerable gifts well. Naturally there are also a few who have used them badly. "It was the first time I had seen such intellectual power in someone who was not good," remembered one person. Another spoke more directly: "I knew before I came here that Fearless Leader is a ——." Such comments were rare, and even these speakers talked appreciatively about certain attributes of their leaders. People of this kind pose few ethically interesting questions, since the wrongness of what they do is obvious. In more established professions, it can be interesting to ask about how the system allows or copes with wrong-doing ("when good doctors go bad," and so on).[16] Bioethics does not seem large or systematic enough to make that question useful. All I can say is that, regrettably but not surprisingly, there are a few bad apples.

What *is* interesting are the moral challenges faced by honorable, fallible human beings who find themselves in leadership positions. One young man described his first years at a center whose leader was well known, career oriented, and indifferent to those who worked under him: "It felt like being followed around by a giant eraser." In many ways he described the older man as decent and principled. But his indifference, combined with his stature and ambition, did the younger man more than a little damage. As the younger man saw it, his every move was lost in a bigger picture dominated by the person I'll call Sam: "Let's ask Sam!" "Sam's really good on this." "Maybe we could get Sam . . ." (reverently). That much was natural, given Sam's prominence; the problem arose with his response. He took on every job and project offered, and did many well, but not all. His subordinates' initiatives and overtures were ignored. Sam would rather do things inadequately than offer the opportunities to others, even to junior faculty whose need he was responsible for recognizing. Good leaders, in contrast, take responsibility for those who work with and under them. A leader has real power over their work lives and their career opportunities. In fairness to them, for the good of the unit and in justice to the field as a whole, a leader should use his or her power consciously and responsibly.

There are other ways that leaders fail that affect everyone, not just the young. A leader whose energies went to staying on good terms with those in power was described by a subordinate as having, in one case, "flat-out lied. He had gone ahead and committed us to something, then presented it to a faculty meeting as if it were an open question." This leader didn't want to offend those with power in the institution. The result was low morale in the unit, and a group of people with little sense of ownership of its activities.

Good leaders do the opposite. They are crucial to a good center. I heard several of them described with appreciation: "He and I not only work well together, we talk about our working relationship." "If anything, she's over-inclusive. Suddenly you're involved in something you knew nothing about." (Involving people without their knowledge is not always a good thing; the speaker just quoted, however, appreciated it. Perhaps the background was that he felt free to withdraw when he lacked time or interest.) Besides the energy and political skills noted above, good leaders are highly principled; as I've argued earlier, the lines between opportunism and opportunity, between expediency and practicality, can be especially fine, but they need to be drawn; an unprincipled ethics center undercuts its own mission. Finally, the good leader is self-aware and self-critical. The need for this is especially acute because, in contrast to the chairs of academic departments or the managers of clinical units, the leader of a bioethics unit is not always subject to review. He or she may hold office indefinitely; normal kinds of checks, balances, and feedback may be missing. In addition, the ambition that leads people to these jobs and to doing parts of those jobs particularly well can also be blinding.

It is an exaggeration, but also instructive, to compare failures here with those for which mountain-climber Jon Krakauer faults himself. "I lived and others died and I didn't do what should have been done. . . . I didn't wait for my teammates . . . an unforgivable lapse."[17] Krakauer survived an Everest climb in which he was a paid customer. Many others died. In retrospect, he believes that he acted like the out-for-himself paying client he was, and not like the we're-in-this-together climber he had always been. When he saw some of the expedition team in trouble, he figured the leaders were in charge and could handle the problems. One moral of this story is that it is not only the official leaders who have an obligation to those who are struggling; we all do. The second, perhaps, is that the more one feels one has "paid" for one's own success, the less likely one is to notice the needs of others.

The Next Generation

There is a glamour to bioethics; people are often fascinated by it but have no idea what the job is really like. Admiring comments—"I want to do what you do"—are common. One undergraduate talked about her future this way: "I might go to med school but I'd like to be employed as an ethicist, not practice medicine. Maybe be on a hospital ethics board. I don't want to be an academic." She didn't understand the basic fact that jobs in this field are rare, and there's no reason to think they will become more common. Everyone in the field with whom I talked understood this and felt a serious obligation to make sure that students learned it. As one put it, "Students have to be mirandized. When someone says, 'I want to do what you do,' I ask more questions, find out their interests. But I don't encourage going into ethics. Find a way to make a living, then see if you can do ethics as part of that. I don't want people to invest a lot of money in education and then not get a job."

Someone else described the precautions his program takes once students have chosen their program: "We tell everyone during the application interview that this is not a degree that will get them a job. Then we make sure they hear it again at least once a semester. If they're already employed professionals, especially in health care, [the degree] may add value to their résumés. They'll probably be qualified to sit on hospital ethics committees, but nobody gets paid to do that. We remind them that only two or three doctors in the whole state are actually paid to do bioethics."

In spite of the general awareness of these facts, the number of new graduate programs and the way some are advertised suggest a kind of blindness. One possible source is obvious: having graduate students is prestigious and rewarding. One of the rewards comes in teaching graduate seminars, during which we can think carefully and at length about bioethics issues, and after which we have more to say, more to publish. (Halfway through writing this book I taught a graduate seminar called "Bioethics as a Practice." It was a chance to do some important basic research and talk through my ideas with intelligent critics.) All across academia, and not just in bioethics, programs are created and protected because of their value for the faculty, rather than for the good of students.

No matter how honestly the employment picture is painted, however, many students enter programs in bioethics. They want to learn and do it, whether or not they can earn a living at it. Some enter programs that require a Ph.D. in religion or in philosophy just for the sake of the bioethics focus. They don't always finish these programs, and even if they do, the fit can be bad. Someone who had gotten a doctorate in religious studies said, "I wouldn't be comfortable in [an academic] religious studies department. [Getting the degree] was just something I had to do." Someone else who had earned his degree in philosophy spoke in disgust of some of the courses he had had to take: "We had to talk about whether there's a light in the room when there's no one there to see it. Give me a break." As it happened, the faculty who had taught him felt as frustrated as he did—"not much philosophy rubbed off on him"—but in other ways there was real affection and respect between this man and his former teachers.

Right now, there are few good paths to recommend to someone interested in bioethics as such. The wisest counsel advises students to study a primary field in which they have real interest and which will provide them with a way to make a living; health care or law are good choices.

Students who choose one of the humanities for their primary field—history, literature, philosophy, qualitative social science—have to be mirandized twice. For thirty years there have been too many graduates of these programs, and the misery of having finished a seven- to ten-year program and being unable to find professional employment cannot be overstated. It is almost worse to get part-time work at poverty-level wages. Consider, for instance, a forty-year-old man with whom I spoke who had a Ph.D. in philosophy and taught "part time" at three institutions spread over a fifty-mile radius. His teaching load was more than double mine and his salary much less than half. He asked bitterly: "Do I even want to stay part of a profession that's so exploitative?"[18] After a few years, such disaffection is not uncommon.

To return to the question, however, suppose that people who have some other way to make a living want training in bioethics. What should that training be like? Some programs are essentially trade schools. A "history" course will include the major court cases in bioethics, but not the history of health care and health policy. A "philosophy" or "foundations" course will look at Beauchamp and Childress's four principles or do a fast day on each of a number of approaches (utilitarianism, pragmatism, "principlism," feminism) but no close reading or argumentation. Other programs are at the other extreme: they offer a traditional and abstract training that does nothing *but* close reading. A bioethicist with whom I spoke found both tracks inadequate. "I wouldn't hire people coming out of the clinical ethics programs, although I would the people from social ethics programs. [I want] people with a humanities background, and an understanding of the larger issues facing society. A background in economics, politics, race, the environment. Most M.A. programs are in philosophy and clinical ethics—that's just not broad enough."

Some may argue that the trade-school model is adequate for those whose only goal is to work with hospital ethics committees and to consult about individual patient-care decisions. Even in these positions, however, to repeat Laurie Zoloth's remark, "It is in part the role of the ethicist to . . . raise the difficult questions that stand just outside of the therapeutic encounter."[19] The vision to see beyond that encounter depends on education, not just training. The most obvious example, perhaps, is the American racial divide that lies behind so many ethics consultations. In addition, many hospital ethics committees are now being asked to take on questions of "organizational ethics," and these involve the topics which the bioethicist quoted above would call social ethics: Should hospitals lobby? Be more responsible about waste disposal? Compete with one another? Advertise? Fight unionization?

Finally, there is no guarantee that clinical ethics consultation, of the kind that has gotten a lot of attention in the field, will continue to be wanted.

My last point is that however students are trained, they need to be treated with respect. I heard mixed reports about this. The following conflicting comments happen to be about the same program. One person described her first week: "All this important info I had so much trouble getting. I didn't even know when classes began, what you have to do. I called the department; they said 'There's a meeting going on right now.' I didn't know about it. 'It was in the mailing.' I never got a mailing. 'It went to your mailbox.' I didn't know I had a mailbox. 'The mailing told you that.'" But someone else, ready to graduate, described one of the professors with admiration: "He gets to know every one of his graduate students so he can write a letter of recommendation that's honest and detailed." Clearly this program does some things well and others poorly. Because these issues are not unique to bioethics, I will not pursue them here, but they need more attention than they get. The long years of conversation and shared action that constitute a virtuous community begin at initiation. The way we treat our young is a good indicator of what we are and what we will be.

Middle Years, Middle Ranks

About halfway through my year of formal interviewing, I began to recognize an experience for which I have not yet found quite the right words. I sometimes saw among those in the middle ranks, people beyond the anxieties of youth but not facing the demands of leadership, a kind of aching knowledge of having been shut out—not from the field as a whole, for these were busy, able people, but from projects for which they would have been particularly competent. One had received a major grant for work in, let's say, the ethics of pediatric research; but her department put together a conference on the topic without including her. Another who had worked for years, and published significantly, on, say, issues in obstetrics was ignored when an institution five miles away launched a major project on the issue. And so on.

Earlier I quoted a young man who felt stalked by a "giant eraser." The metaphor would be apt for many of the people with whom I spoke: "I wrote the report, but it was published under a committee's name." "I was told I'd have forty minutes to describe my work, then found out I'd been cut to ten. The organizer had never even asked what I would be presenting."

Several speakers used the word "integrity" to describe what they felt had been lost. Sometimes they were referring to accountability, which—for instance—is missing in an anonymous group report; but most meant by integrity a responsibility to themselves. It can be wrong to let one's work be treated as if it did not matter.

The belief that one has been overlooked, and undervalued, is a common source of suffering for midcareer and senior professionals. I heard many examples of it, rarely bitter or petulant, usually sad, surprised, perhaps defeated. The causes varied as the circumstances did. One tried to analyze the leader of her unit: "What's so amazing is his obtuseness. He doesn't do this on purpose; he just doesn't see. Of course he does deliberately schmooze with whoever's got power—he sees them, all right." From what I could see, her assessment seemed accurate. But the issue extends beyond any particular unit. In our broader interactions, our professional lives regionally and nationally, we can fail one another and fail the work. We overlook one another for many reasons: because we like the comfort of the familiar, especially when we are overworked; because we are ignorant, sometimes culpably so; because we need to please our closest disciplinary colleagues; because we are arrogant. Sometimes the cause is the personal fault of a conference organizer, book editor, grant bestower, or unit leader; sometimes it is just the inevitable arbitrariness when scarce resources are distributed. Sometimes the cause is heedlessness. Blinders like the ones I've just described are especially effective when they are worn by people who are ambitious or in a hurry, and that describes most of us. As a newcomer to the field remarked of her first conference, "At every break everyone ran to the phone."

And that brings me to my next topic.

The Practice as a Community

The practice itself, at a national and international level, is a kind of community: its customs and institutions can be respectful or disrespectful, and they can foster or impede good work. They can make possible the long, respectful, challenging collaboration that leads to moral growth, or they can fail to do so.

The emergence during the 1990s of a single dominant professional association, the American Society for Bioethics and Humanities (ASBH), has done a lot to facilitate productive interaction. (One person compared the small pre-ASBH associations to "separate sandboxes" and to "people drawing fences.")[20] Since the merger, a national task force has drawn up a set of standard competencies for doing bioethics consultation and education, which ASBH adopted and promulgated.[21] This is a step toward a responsible community, one that holds its members accountable and tries to prevent its name from being used by those on the make. (It is still possible for anyone to "hang out a shingle" and try to make money as an ethicist, but the initiatives of ASBH make that less likely.) The association has also taken steps to help young bioethicists, in particular a web-available booklet with practical guidance on publishing.[22]

There are less formal, but still public and structured, ways in which members of the broader community influence one another. On the electronic list hosted by the Medical College of Wisconsin, for instance, contributors question various practices, among them graduate programs that advertise "exciting employment opportunities."

Other moral challenges to our community as bioethicists are subtler, more ambiguous; some of our customs both promote morally challenging collaboration and impede it. I will discuss three in particular. One is the role that personal and institutional connections play; another is the entrenched American emphasis on competition and evaluation; the third concerns the effects of spending an ever-increasing number of hours each week working.

Let me introduce the first of these points by citing one of my interviewees, who told me that every job she had ever had, she had gotten through connections. In particular, she identified informal apprenticeships as having been crucial to her career. I asked someone else how he came to be involved in a large and interesting grant-funded project; he explained that a funding agency had approached "the grand old names" in his institution, who were not themselves interested—"but we have these young people who might be." This group of senior scholars was willing to share the opportunities that came their way, to nurture their younger associates; in that sense the story is exemplary. But it also illustrates some of the basic ways in which position and connections matter. This is so self-evident that when the directorship of the Hastings Center changed hands a few years ago, someone remarked un-self-consciously, "Great. We know him, we've worked with him for years, now we'll have an in."

There are good reasons for such connections to matter. I learned that lesson when I was putting together panels for various professional conferences. At first I diligently tried to recruit from a wide set of people, not just from those I knew. Once I posted an inquiry on an electronic list, carefully pruned the names that came forward, sought out recommendations from others (especially from people whose judgment I trusted—although that, too, perpetuates position and privilege). Nevertheless the result was disastrous, the worst panel I've ever convened. After that I became more cautious, more likely to choose from people I knew personally.

So the power of affiliation is inevitable, necessary, and an aid to excellence. But it can be unfair and wasteful as well, because it can blind us to the abilities of those outside our own circles. I once escorted a visiting speaker around the MSU campus. He remarked about how much he'd enjoyed his lecture tour: "I've found out there are lots of really smart people around the country; they're not all in Boston." He was not being ironic; he was genuinely congratulating himself on his "discovery." He was not alone in his insularity, and I am not alone in noting how infrequently it is challenged. "Insularity" here is not a matter of geography but of class, in its academic form. Class in America is virtually invisible. In spite of the mantra "race, class, and gender," class enters the discussion only when it takes the form of poverty, and not always then. Differential privilege, reinforced and reproduced, in the graded ranks above destitution goes almost entirely uncriticized. The power of one's institutional connections functions within academia the way socioeconomic status functions outside it, and yet few of us recognize being privileged in either of these ways as anything except the result of merit. In some ways such power is less within bioethics than in more established fields, simply because it is newer and fresher. In other ways it seems to me worse, since standards of excellence are unclear and objective evaluation is harder.

There is no doubt, however, that throughout the disciplines institutional status is hugely important. One young physician told me how much she had loved her medical education (in itself an astonishing remark; medical school is grueling and unpleasant, on the whole). As she described it, she and her fellow students had far more contact with senior physicians, and ongoing relationships with patients, than most schools provide. But the school was unranked, without prestige. She applied for residencies at several prestigious institutions, and faced constant suspicion because of where she had gone to medical school. During one interview, she said, "I almost walked out, they showed so much disdain." But she stayed, and got the residency; eventually she realized that she knew "as much—sometimes lots more—as residents from more prestigious schools." In spite of that, it was the reputation of the residency and not the unrecognized excellence of her medical school that helped her get the (quite prestigious) fellowship she now held.

We narrow our circles not only by personal acquaintance and institutional connections but also by disciplinary boundaries. A social scientist said to me, "I always feel I'm 'out there,' 'marginal,' because I'm [a social scientist]. All these powerful networks. . . ." In the chapter on the "languages" of bioethics I looked more closely at problems in working across disciplines, as I will in the next chapter, on interdisciplinarity. Here I just want to add disciplinary myopia to the list of blinders that can interfere with good work and mutual respect.

There is one famous, ongoing story within the field that illustrates the power of connection and some of the paradoxes that entangle efforts to correct it. Years ago, the legend goes, Art Caplan, one of the most publicly visible bioethicists in the country, began what has come to be called "summer camp." Caplan is said to have drawn up the invitation list from his rolodex. The group was small and select (or at least selected) and the conference was different from the standard academic conference. Everyone was together for every session and every meal, and no papers were read, but ideas were circulated. Once, an early participant told me, the group called one of their own to task for the way he was dealing with the media. Soon (or perhaps from the very beginning), families were invited along, and a good part of each day was free for recreation. Usually the sessions were held in the country, rather than in the city. "Summer camp" became an institution, sponsored by different people each year in a different part of the country. Its invitation list gradually grew, partly because of some uneasiness about its elitism, and the list now is considerably larger, but the event is by no means open to all. Some people refuse to go on principle. (One prominent person new to the field told me he's eager to get an invitation so that he can decline it.)

One year, it was reported to me, a group of young people began a rival summer camp for junior scholars "and it was great for networking," but in the end "it got co-opted." (I am not sure how.) Another year a sociologist got himself invited to the "senior" camp, and took notes—and reports that he was virtually evicted when his note taking was noticed. Sometimes summer camp is held at expensive, luxury locations. The melding of vacation with work yields something like what is called in game theory, and in moral theory, a prisoners' dilemma. That is, each of us individually can say that for professional growth (both intrinsically, in what we learn, and extrinsically, by the connections we make), summer camp is important. Therefore it is legitimate to ask one's institution to meet the costs. The vacation aspect may play no part in one's desire to attend; summer camp might even be held in a setting one does not enjoy. But the net result is that private enjoyment is sometimes paid for by professional funds: health care in some cases, taxpayer funds in others.[23] Even here, however, some would point out how fruitful they find the conversations during the recreational hours, and they could argue that a morally useful community will cherish the chance for extended, exploratory conversation.

So there are many moral ironies connected with summer camp,

and at least one of them, the question of exclusivity, is the subject of regular breast beating among participants. The most interesting issue here may be that some feel it has lost its original usefulness as it has lost its original character. It is no longer small enough or private enough for anyone to be "brought to task" by the group. The conscious formation of group standards that sometimes happened in the beginning is rarer now. Broadening and democratization, morally good as they are, have also had moral costs. A moral community somehow has to navigate a path through all this.

My second focus in discussing bioethics as a national community concerns competition and evaluation; here the picture is similarly paradoxical. Americans tend to assume that both are necessary conditions for excellence. Most academics, I would guess, understand the ideological biases of this assumption, and know that cooperation is at least as important. Competition plays little part in the communities identified by Colby and Damon, and by Cherniss. Alfie Kohn, an organizational behavior theorist, points out that competition *by definition* requires some to fail so that others can succeed, and found it counterproductive, at least within firms. He argues, instead, for "positive interdependence," in which team members "work for the same goal, use the same resources, and receive the same rewards. The shared group identity that results is a powerful motivator because one person can succeed only if the others succeed, too."[24] Interestingly, he emphasizes that differences of opinion will continue to exist, a key finding of Colby and Damon as well.

It is difficult to extrapolate from this to ideals for the profession as a whole. We all feel the effects as the field gains attention and respect, or conversely when it becomes cheapened in some way. We all benefit when ethics and the humanities become standard within the education of doctors, nurses, and therapists; when Congress and funding agencies allocate money for the kind of work we do. Not all of those rewards can be shared equally, however. Funding and fellowships must go to some applicants rather than to others. Not everything can or should be published. In other words, both cooperation and competition are inevitable.

At a minimum, however, we should be realistic in our thinking about competition and evaluation, and try to demythologize them. One way to do that is to remind ourselves regularly of the factors besides excellence and hard work that contribute to success and failure. I've already pointed out the effects of connection (personal, professional, and institutional) and some of the ways race, class, and gender can make a difference. More can be said about this latter point. Gender expectations generally make women into the helpmates of men. One woman with whom I talked described a male colleague whose career was aided immeasurably by his wife, whom I'll call Susan. He dictated his articles, she wrote them up and checked the references. They traveled to conferences together; she drove, so that he could read or dictate while riding. The woman who noted all this remarked ruefully, "I don't have a Susan. Besides, I have to BE a Susan for a lot of people." Class works itself out in similar ways. Someone described the work life of a nationally known figure: "An assistant follows her to every talk, tapes it, transcribes it, prints it out, and edits it. One talk, one article. . . ." A more significant form of this kind of advantage is the help some institutions provide for the writing of grant applications; some, in fact, employ people just to write grant proposals. Nothing succeeds like success.

Achievement in any field, therefore, depends on more than effort and ability, important as those are. It's important to keep this fact in mind, so that newcomers understand the landscape and veterans can consider leveling it a bit. A certain rather dramatic parallel has startled some people into thinking afresh, and I offer it here: Most readers know that admission to medical school is highly competitive and inescapably arbitrary, since there are few reliable indicators of what will make someone a good physician. (The practice of medicine demands much more than the ability to learn facts, which is what standardized examinations test for.) Against that background, the Dutch chose until very recently to make admission to medical school a matter of chance. I exaggerate; to be eligible, a student must have graduated from what we would call a highly demanding college preparatory program. But with that diploma in hand, all one needed to do was apply to medical school. Successful applicants were chosen from the pool at random. The system manifested candor about the impossibility of choosing accurately as well as a commitment to egalitarianism, without sacrificing quality.[25]

I don't suggest any close parallel in bioethics. I tell the story because it is so counter to American assumptions that it might jolt our thinking into new paths.

The final aspect of our national professional culture that deserves attention is the amount of time that work takes. This is true, not just in bioethics, but in most fields, and it is tied in part to the emphasis on competition and evaluation. Part of this picture, both cause and result, is that far too much is written and published: no one can begin to read everything (even in a narrow subspecialty), and much of what is published is not worth reading. If people are rewarded for publishing in professional journals, more professional journals will arise—as they have.

There is another sense in which competition and evaluation drain time from useful work. As I wrote this chapter, for instance, I received a phone call from a friend asking me to write a letter for his promotion file. This will demand reading the materials he sends me as well as composing a two- or three-page, detailed recommendation. A journal has asked me to "referee" an article—to read it and make a recommendation, supported by reasons, about whether to publish it. My own institution has asked me to make a summary of all my professional accomplishments for the past year (something we seem to be asked to do every few months, in different format for different purposes). Since it is not clear whether the "past year" means the calendar year or the academic year, our secretary is making inquiries. A colleague from another unit, with whom I am working on a paper, has just returned from spending days, nearly a week, serving on a panel evaluating grant proposals. I described in another chapter a scientist who spends 50 percent of her time writing grant applications and evaluating the applications of others. The fraction in bioethics is lower, but it is still significant. None of this is directly productive time. It is all evaluative. We sort, rank, and grade when most of us would prefer to be reading, writing, and teaching.

These and other factors contribute to work lives that are frantic. A young physician just finished with residency said to me, "I so badly wanted to stop. . . . I'm getting my life back together. . . . I never want to be this devoted to any job. . . . People don't realize until they stop. . . ." She was speaking of her years in medical training, not in bioethics and the humanities, and residency is known for its almost inhuman demands. Yet by most accounts almost everyone's work life is too pressured today.[26] During sabbaticals, academics realize what this young doctor realized when her residency was over, that life can be different: it is possible to work well and still have time for the rest of life. Significant decisions get made after these periods of peace: to resign, to end a marriage, to retire early for the sake of a marriage. But not everyone gets a sabbatical, not everyone takes one, and most of us who do must simply return to the rat race when it is over.

The one concrete suggestion I can make here is to encourage sabbaticals and other forms of leave; they are unique in providing time for reflection, and reflection is necessary if we are to take charge of the customs, habits, and structures that shape our separate and shared professional lives. I tend to believe most of us would do well to read more widely and write less, to take on fewer projects but do them more thoroughly. It's difficult to make such choices on one's own, however; a sort of prisoner's dilemma operates here as well as other places: what would be best for the whole group is suicidal when undertaken alone.

Drawing any other moral about the larger community being formed in bioethics and the humanities is hard. Much of what I have described is unavoidable, much of it understandable, some of it justifiable. But it is important at least to be candid and searching about what is happening. The pursuit of individual professional success will not by itself produce the common goods we hope for. A virtuous community provides a structure in which supportive but probing conversation is constant: it nurtures its young, welcomes a variety of voices and energies, and shares its resources and opportunities. It is self-conscious, understanding its own structures and implicit values well enough to use rather than be used by them. Bioethics at the beginning of the twenty-first century, filled with energy and possibility, is well situated to reflect on the communities it is creating.

One feature of these communities is so distinctive and so important that I've reserved a separate chapter for it: The work we do as bioethicists is by its very nature interdisciplinary. We are also by definition engaged in an intellectual pursuit, in strengthening moral understanding in ourselves and others. Intellectual virtue takes a distinctive shape in an interdisciplinary context. And this is the topic to which we will now turn.

10 Intellectual Virtue and Interdisciplinary Work

Although work in the medical humanities is not *just* intellectual—what makes it so interesting and so puzzling is the fact that it is practical as well—the work must be intellectual at its core or turn into something else entirely, into mediation or activism or therapy. (I'm speaking of the field, not of each individual within it. Not every member of, say, a hospital ethics committee will write or even read the academic literature. But they use it, even if sometimes unknowingly.) The moral growth in which we participate entails reasoning and reflection, about clinical, policy, and research decisions, and about the life-worlds of those we hope to work for and with. It follows that a virtuous community doing bioethics must have intellectual as well as moral virtue. Intellectual virtue is especially challenged in interdisciplinary work.

Intellectual virtues are those qualities, in individuals and in communities, that promote growth in knowledge and understanding. Several of the people with whom I talked mentioned intellectual virtues—"being analytical," "honesty and clarity"—as essential for the field. Many had strong opinions about the intellectual life provided by the profession.

Comments from the Field

Some people, as I described in the "territory" allegory, move to bioethics because of intellectual hunger. Even those who were not dissatisfied in their original field can find special intellectual pleasure in their new work. Said one person, "I was a competent nurse, but bioethics was where my heart was. I just fell in love with . . . the chance to be creative; to hear really different points of view; to question everything." Another spoke of a fellowship year: "This has been the most stimulating year of my life—the readings, the discussions. Thinking about a lot of things new to me, and interacting with people who have interesting things to say. . . . Whenever I tell people what I'm doing they say, 'Oh, I wish I could do that.' . . . The issues are challenging and important; I respond to them, and I want to help shape the response they get."

Others are just as strikingly dissatisfied. "Bioethics has flowered like an overfertilized plant; it needs work at the roots. . . . There's a rush toward outreach, toward credentialing ethics committee members . . . while the literature of the field is simply boring." "I just found bioethics too thin. In religious studies [in contrast] I can talk about things like my sense that fragility has beauty." "It's just too shallow." For what it's worth, those who seemed most dissatisfied often found their real homes in one of the traditional humanities; and some who found the *literature* of bioethics superficial were quite happy in the work, which involves so much more than text. Those who reported "falling in love" with the field often did so first during an undergraduate course. Most colleges and universities now have an undergraduate course in bioethics, often taught in philosophy or religion, usually in high demand and well received. The courses are fairly standard now, the same in essentials across the country, supported by a shelf of competing textbooks that deal with the same basic topics: "end-of-life" issues (terminating treatment, euthanasia, assisted suicide), research on human subjects, justice and access to health care, and so on. Popularity with students does not guarantee that a course has depth, of course. But bioethics has been well received for decades at many different campuses, and its staying power suggests that undergraduate courses are challenging and satisfying. At this level the field works well by virtually all accounts.

At other levels, obviously, opinions differ. My own opinion is that the intrinsic practicality of the field has to affect what counts as good work, and that looking for analogs to great literature, philosophy, or sociological theory may be a mistaken quest. (Although certainly important works from these fields find permanent homes within bioethics. Tolstoy's "The Death of Ivan Ilyich," Paul Starr's *The Social Transformation of American Medicine,* and Charles Bosk's

Forgive and Remember come to mind.[1] But none was written *as* bioethics; each fits within a more established genre or a discipline.) There are classic works within bioethics: Henry Beecher's list of questionable research projects helps us mark a beginning, and progress.[2] The reports of the National Commission and the President's Commission provide authoritative statements of reasoned consensus on practical ethical issues.[3] But these are important for their historical role or their usefulness. They are not the kind of thing that will be read a century from now for their literary or intellectual satisfactions.

The work I find most satisfying currently is of the same sort, clearly written and argued, and useful for a significant set of specific practical issues: Joanne Lynn and James Childress on whether patients must always be given food and water, Alan Meisel on legal myths about terminating life support.[4] The best work of this kind must be empirically informed and well reasoned, and the satisfactions it offers are specific. The pleasures of bioethics, for now, are often those of *praxis* rather than of text alone or action alone: in good bioethics, theory reflects upon practice, and practice informs theory. Over time that may change, and some writing may emerge that will stand on its own.[5]

Praxis is unavoidably interdisciplinary, for the same reasons that no discipline alone is adequate for moral growth. The a priori abstractions of philosophy are pointless or dangerous unless one understands how the world actually works; the patterns of power and systems of meaning identified by social science do not by themselves answer questions about justice and injustice. In addition, these disciplinary boundaries are not so neat as I make them sound, a fact that makes the need to work together still greater.

So the thinness that some find in bioethics may result from a mistaken quest. Bioethics should not (at least yet) be expected to yield the intellectual greatness of the best literature, philosophy, history, and other scholarship. Yet criticism of bioethics may be justified on other grounds. In particular, the interdisciplinarity of our work presents special challenges.

The Promise, and the Pitfalls, of Interdisciplinary Work

Fittingly, the term "interdisciplinary" itself has no single meaning. One person who works across disciplines spoke of her frustration in this regard: "Funding agencies think of interdisciplinarity as a historian reading literature, or vice versa. Where I think it should be collaborative work that results in a whole new thing." It is not just in bioethics that the meaning of the term is unclear. The terms "interdisciplinary," "multi-disciplinary," even "trans-disciplinary" occur but are constantly contested. Obviously each assumes a contrast with "discipline," but that term, too, is more fluid than one might think. What we now consider the established disciplines are in some cases only a few decades old. Even the laboratory sciences rearrange themselves, as old fields become exhausted (there are no new findings in anatomy) and new ones arise (like molecular biology). Higher education has arranged itself differently at different times and in different countries, and for a variety of reasons.[6] But even though the traditional disciplines are far from Platonic Forms, they do have something usually lacking in interdisciplinary work: forms of inquiry with well-understood, shared norms. Without these, research is filled with excitement but also with risk.

One commentator believes that the term "interdisciplinary" first emerged within grant-giving agencies in the 1970s "as a code word for politically or theoretically adventurous work (feminism . . . work in media and mass culture . . . Marxist and psychoanalytic criticism). The term had a useful function, then, in making this new work look professionally respectable and safe."[7] Today, he believes, the term has become tame, institutionalized, and uninteresting, a euphemism "that permits us to feel good about what we do and to avoid thinking about it too precisely." He longs instead for "forms of 'indiscipline,' of turbulence or incoherence at the inner and outer boundaries of disciplines. If a discipline is a way of insuring the continuity of a set of collective practices (technical, social, professional, etc.) 'indiscipline' is a moment of breakage or rupture."[8]

These phrases express an idealism that remains common, a longing for the transcending of boundaries, the critique of convention, and the revelation of new complexities.[9]

The term "interdisciplinarity," then, is only loosely defined, has been given a variety of theoretical analyses, and is viewed by those involved in it with a mixture of cynicism and excitement. The mixture is probably healthy, and the looseness of definition necessary. As one writer comments, to think that some absolute criterion of interdisciplinarity exists is to treat it as a discipline—which is exactly what it is not.[10]

I will use the phrase "interdisciplinary work" simply, to refer to work that involves systematic study but does not fit within the boundaries of what we now think of as the traditional disciplines. Whatever language and theoretical constructs we use to understand interdisciplinarity, probably the most fundamental point is that "the *real* problems of society do not happen in discipline-shaped blocks."[11] An analogy I like for its modesty and its caution is that of bricks and mortar, an image its author calls "useful but imperfect. . . . Bricks need mortar to join them, which must sometimes span a fair gap. Bricks may also have to be cut to fit, or may be at risk of failure when used other than their makers intended."[12]

One may hope that the new work will eventually do more than combine blocks from old fields, but in the beginning that is often all it can do. As a result, the author whom I just quoted continues, we need to be prepared to criticize "the wholesale, uncritical importation of small chunks of analysis . . . from their disciplinary origins" into a new set of projects.[13] We should notice that the imported tools and information will probably be simplified, and possibly oversimplified. After that things can get worse, as uncritical repetition makes certain claims into conventional wisdom. Significant problems can arise when, say, philosophers and nurses and theologians quote a nonhistorian on some point of history, and then quote one another, *ad infinitum.* Bob Baker has made this point several times on an electronic listserve: "My standard complaint about bioethics scholarship is that it tends to be a mile wide and millimeter deep with respect to history. Ranging back to the Hippocratic

Oath and Nazi genocide, without reflecting in depth on either. . . . *They need to read historians proper and integrate the current historical literature into bioethics*"[14] (emphasis added).

I first recognized this issue when, reviewing a book by a biologist, I realized that she was quoting studies from the social sciences uncritically, having no tools with which to evaluate them. I didn't have those tools, either, but—in that instance—I knew that I didn't. A few years later, a lecture by an epidemiologist jolted me into seeing that I had been doing with medical studies what the biologist had done with social science: quoting results whose methods I could not evaluate.

Many reflective interdisciplinary scholars have noted the danger, intrinsic to their work, of "scavenging," "ransacking," and "questionable eclecticism."[15] To some extent this echoes MacIntyre's point, that only those who understand a practice in its own terms can judge the quality of its internal accomplishments. (However, sometimes this criterion is too strong. Research design and statistical methods can be largely the same across many fields.) The responsible alternative to "scavenging" is to recognize what one does not know and either cease the citing, enlist an expert, or learn some basics.

This problem is a fundamental danger to good work, in the sense of a good product. It also interferes with good work in the sense of satisfying activity. For instance, I was once part of a committee judging a competition for best student paper. When we sat down, we found that the same paper had been judged best by some and worst by others. Although we eventually reached a consensus, no sudden light of conversion occurred; we each stretched a little and conceded a lot. If the lack of shared standards makes judges uncomfortable, it can leave those judged baffled and frustrated. Referees' comments on work submitted for publication in bioethics can be opaque. In the case of one of my own papers, language ordinary to me ("overdetermined," "pathologizing") was jargon to a referee. In addition, he wanted a degree of precision in my use of epidemiological data that at the time I thought was quibbling—I was only using the material to illustrate general, *philosophical* points. But I came to understand that he was right in wanting what was, after all, simple accuracy.

I have attributed these problems to a lack of shared standards; more deeply, what we lack is a common methodology. The long training that makes one a professional within any field includes tacit as well as explicit learning, about ways of handling test tubes, data, arguments, or texts. One acquires a set of assumptions; learns habits of mind that make certain things salient, others background; learns about past inquiries and their successes and failures. People who decide to work together across such distances (and partly because of them, since so much can be learned on the journey) confront these differences and often resolve them. A harder project is communicating with people who do not make the journey.

This project is not about responsible inquiry but about talking with people who live in different conceptual worlds. Such conversation is essential to bioethics, a field that wants to affect what people do. In some sense many of us have learned to bridge some of the gaps, to teach and speak to clinicians, scientists, and policy makers. Writing for them, however, is harder; they have habits of mind and of reading quite different from our own. Bioethicists do publish in clinical, scientific, or professional journals, but the results are hard to sort out. Within clinical bioethics, publication in the best medical journals (*JAMA*, *New England Journal of Medicine*, and a few others) is an important way for ethicists to communicate with one another. The same does not seem to be true of ethics publication in science and policy journals. But even in clinical bioethics, it is unclear whether nonethicist doctors read such pieces. A physician newly interested in ethics said to me, "I thought of [ethics articles] as soft. And I didn't have much time."

Gladys White made the same comment about scientists: "The best bioethics literature is unread by them." She finds the results unfortunate: innovative procedures in, say, reproductive technologies become standard practice before anyone has a chance to raise ethical questions. When I spoke with her she was executive director of the National Advisory Board on Ethics in Reproduction (NABER), an independently funded agency whose purpose was to raise ethical questions early.[16] White thought that a real solution would demand a "finer mesh" between innovation and practice, so that the people doing research and the people doing ethics were talking and thinking together from the beginning.

Having more conversations earlier would have benefits beyond those White identified, benefits throughout bioethics, because we understand our fundamental questions better when we see them asked in different contexts. (What does a right to refuse treatment mean when the patient is an adolescent, or a trauma victim, or a schizophrenic whose disease is controlled by medication?) Almost always the new understanding is instructive beyond the field where it first occurs.[17] Seeing old questions in new ways demands attentive involvement: observing, listening to, reading about, and talking with those engaged in the practice. It involves, in other words, praxis rather than mere application.

However, the application model remains common, and many of those with whom I talked found the results disappointing. One called the philosophy in articles published in medical journals "transparently weak." Another used the "search and replace" metaphor I mentioned earlier: "Informed consent in dermatology . . . in neurosurgery . . . in otolaryngology. . . ." I remembered these comments when I was asked to evaluate the work of someone applying for promotion. As I read I found that I could sink immediately into the articles written for philosophy journals (hardly a surprise) and had no trouble reading those he published in the *Hastings Center Report*. But those written for the subspecialty journals—the "informed consent for otolaryngology" kind of thing—were dry as pasteboard. The pidgin metaphor I developed earlier in this book explains part of the dullness: the articles had to fit ethics content into a scientific format. Another factor is that bioethics basics are still unknown to some people, and the articles had to serve as primers. But I continue to wonder if those for whom the material is written actually read it. I hope we can find a better way to communicate.

So interdisciplinary work can be shallow; it can fail to achieve its promise. At its best, however, it is like bringing lamps into a cave. Bringing light, I have argued, demands working together rather than observing and instructing one another from our disciplinary corners. Working together in turn demands intellectual virtue. On this subject there is a great deal of new work within philosophy.

The Nature of Intellectual Virtue

Remember that the concept of virtue is teleological: virtues are excellences that either in themselves or through their results contribute to human good. One of the major purposes of bioethics is deepening our moral understanding. This quest, so much broader than the common misperception of our goals as a set of answers, would have fit badly into much of twentieth-century epistemology. For the most part that field understood knowledge as something like information and dealt with the justification of knowledge claims that were atomistic in every sense: one knower, one moment in time, one proposition.[18] Although this project is far from exhausted, many epistemologists now understand their subject matter differently. In the words of Jonathan L. Kvanvig, "Knowledge comes in bodies and information in chunks [which in turn] come structured, with paths or routes built in between compartments in the body or chunk."[19] It follows that we should move away "from the assumption of an ordinary, competent adult isolated from the flow of time in studious contemplation of evidence, warrant principles, and argument forms having individual propositions as their (warranted) conclusions. . . ."[20] Instead, we form communities into which others are socialized, which seek knowledge for a variety of "plans, practices, rituals and the like" in the service of those practical and theoretical purposes that help define us.[21]

Into this richer, although rougher, picture of knowledge, the pursuit of moral understanding fits well, as does the importance of interdisciplinary work. What virtues, then, help us in our pursuit? There has been a recent outpouring of philosophical work on the general question of intellectual virtue, work which draws from Aristotle but is very different from his. In turn my own discussion draws from the recent work but differs (in purpose).[22] Aristotle thought of intellectual virtue and moral virtue as quite distinct. The former were qualities (or better, excellent activities) of the mind; the latter excellences of character. Linda Zagzebski argues for a different categorization. "Intellectual virtues are best viewed as forms of moral virtue."[23] This is so because gaining understanding is *an activity.* We who inquire, who learn, and who construct a coherent whole out of what we learn, are in all these activities doers, capable of self-reflection, and as such subject to moral evaluation.

Zagzebski points out a number of similarities between intellectual and moral virtue. Both "require training through the imitation of virtuous persons and practice in acting virtuously. Both also involve handling certain feelings and acquiring the ability to *like* acting virtuously."[24] Both are also matters of degree. In addition to these formal similarities between moral and intellectual virtue, there are causal relationships. Moral virtue, and its lack, help determine whether we attain intellectual virtue. "Envy, pride, and the urge to reinforce prejudices can easily inhibit the acquisition of intellectual virtues. A person without sufficient self-respect and an inordinate need to be liked by others may tend to intellectual conformity. An egoistic person will want to get his way, and this includes wanting to be right."[25] As a result he will resist evidence suggesting that he is wrong. His energies will go into debate rather than into the pursuit of truth. "He has, then, intellectual failings resulting from a moral vice."[26] Interdisciplinary work can fail in this way, as can any academic pursuit. About the special needs of interdisciplinary practice I will say more later.

If moral virtue contributes to intellectual virtue, the reverse is also true. In order to be fair, or generous, or courageous, or compassionate, Zagzebski argues, one must know what the world is really like. Otherwise an attempt at, say, being compassionate could end up being cruel, an effort to be brave might instead be foolhardy, and so on.

Finally, in attaining any virtue there is a significant amount of luck involved. Since all virtues are learned, Kvanvig notes, "there is the luck of latching onto the right exemplars" (teachers, models, mentors) as well as the good fortune in having "sufficiently adequate native equipment" to be able to emulate these exemplars and grow.[27] It follows that communities and not just individuals can have, or lack, intellectual virtue. It exists in communities because we learn skills of inquiry (or "the know-how . . . at the heart of the cognitive life") from one another in organized ways.[28] This basic fact of human life gives rise to responsibilities, often communal ones.

Some Individual and Communal Intellectual Virtues

Intellectual virtues include courage, conscientiousness, humility, impartiality, honesty, integrity, and self-reflection. (There is of course no simple complete list.) The nature and significance of most of these qualities are obvious, and so is their relevance to interdisciplinary work. Conscientiousness, for instance, is crucial when one is working on unfamiliar ground. A few of these virtues, however, deserve further comment.

Courage, for example, may look surprising on this list. But Lorraine Code reminds us that "it requires courage to become reconciled to dealing in areas where certainty is not possible, where the subject matter is amorphous and, to a great extent, unmanageable. . . . On all sides, one is faced with the fact of one's own fallibility, of human fallibility in general, and of the need to acknowledge this fallibility if better understanding is to be achieved."[29] In bioethics we are constantly in such situations. Several people mentioned other kinds of risk that our work demands we face: "It takes guts to walk into a clinic not being a doctor." "Should we be interacting with the press? Definitely yes. You've got to take a stand. . . . These things affect people; we should allow ourselves to be quoted, skewered, humiliated . . . to be tested, pushed beyond our shared assumptions."

Walking into a clinic, talking with the press, teaching in a medical school (about which I will say more later)—obviously these things take courage. But is courage an *intellectual* virtue in these contexts? Yes: we enter these arenas not as missionaries with something to give, but as explorers, or perhaps traders, with much to receive and in particular much to learn. Understanding the life-worlds of others is central to our job, and it can only be accomplished by putting oneself in uncomfortable situations.

Honesty may seem a more obvious intellectual virtue. But Zagzebski reminds us that it is important to see honesty as more than a matter of avoiding lies. "An honest person is *careful* with the truth. She respects it and does her best to find it out, to preserve it, and to communicate it in a way that permits the hearer" to be justified in accepting it.[30] Understood in this way, honesty entails conscientiousness, which in turn demands attention to the limits on one's claims to know. The relevance of this to "scavenging" in interdisciplinary work should be obvious. Conscientiousness, however, demands attention not just to evidence and logic but also to oneself. As Code remarks, it is "precisely because self tends to obtrude so insistently in all . . . attempts to be 'objective,' that self-knowledge is crucial."[31] Understanding oneself is not an easy task, either for individuals or a community. (My last chapter tried to increase bioethics' understanding of its own community.)

Intellectual virtue depends fundamentally on what one values and why. Code argues that "an epistemic community will be strong in intellectual virtue only if good knowing is valued as a condition of human flourishing, whether individually or collectively."[32] Knowledge cannot be seen simply as an instrumental good. This is a subtler point than it might seem. It is not just that knowledge should be valued for more than its ability to earn us money and status, issues I've talked about before; knowledge should also be valued for more than what it contributes to the good of patients and the public. If Code is right, a community with intellectual virtue will honor the activities of learning because they are part of a good individual and common life. These two purposes (knowledge for the sake of accomplishing good and knowledge as enriching the knower) are rarely in tension with one another; in fact they are hard to separate. Aristotle reminded us that human beings by nature desire to know, and cited as evidence the delight we take in our senses.[33]

Reliability is a communal intellectual virtue. In general, people should be justified in placing their trust in one another in a common pursuit of understanding, trust not only that they will not be deceived and misled, but that they will not be abandoned if they join projects that demand years to complete.[34]

Mentoring of the young is another communal intellectual virtue. Since it is not just rules from a book but embodied technique that must be mastered, those new to the field must have teachers and exemplars. One of the people with whom I talked mentioned advising her students to skim certain kinds of material; "nobody ever tells them that." Another advised students to choose a research topic by doing a literature search, looking among other things for how many articles had been written on a topic. If there were several hundred, he said, one should abandon the subject; if there were ten or twenty, look for what is missing. Without this kind of practical advice, reading and research are impossible tasks. Obviously, though, the search for knowledge is more than a matter of technique, and the young need more than access to smart, articulate people. What is needed is "an interpersonal relationship with [exemplars] . . . that is at the heart of the life-long endeavor of finding the truth."[35]

Besides an orientation toward truth and enduring relationships among members, a community fosters (or impedes) intellectual virtue through custom and structure. The scientific community has traced some of its recent problems with dishonesty—with falsifying data, for instance—to the intense pressure to succeed. The pressure also encourages a proliferation of minor publications, to which some universities have responded by changing the way they evaluate candidates for promotion: no matter how many articles someone has published, only five may be submitted with a tenure application, and only ten with an application for promotion to full professor.

I have dealt with most of these intellectual virtues quickly. It will be useful to look in more detail at one that melds individual virtue, community structures, and an intellectual life in surprising ways.

An Example: Humility, the Pursuit of Understanding, and the "Audit Culture"

Humility as it is usually understood requires that we not overestimate ourselves and that we keep our limitations clearly in mind. The need for all this in an interdisciplinary field, and one which deals with the sufferings of others, is probably self-evident. Yet there is a less obvious face to humility, which is worth attention.

This virtue has been treated variously at various times. During the Middle Ages the Christian tradition tied humility to self-abnegation and even self-contempt in a way that is impossible to endorse today. But contemporary religious and philosophical treatments make what seems to me a different kind of error: they characterize humility as something like good vision, the ability to see one's own status and abilities accurately, with a special caution about overestimating them. I would argue that this ability to see clearly is the *result* of the virtue rather than the virtue itself. The virtue itself includes—crucially includes—an emotional disposition which makes this accurate self-knowledge possible.

To see this, we need to think about what self-knowledge is like and how it is attained. Learning about oneself is quite different from, say, learning biology or law. These subjects can be taught, but self-knowledge can only be fostered and encouraged. Others—teacher, therapist, enemy, friend—can help us learn about ourselves, but they cannot simply *tell* us about ourselves. They provide information, by what they say and how they respond to us, which we may or may not be able to absorb. I would define humility as the easy recognition of one's failings and successes—not just recognizing them, but doing so in a way that allows a fruitful rather than a destructive response. The ability to do this depends upon a basic kindness toward oneself without which self-knowledge can seem too frightening to allow. With such kindness, however, it is possible to acknowledge even painfully deep deficiencies in a way that increases inner harmony and strength.

This describes humility as a general moral virtue. Its role as an intellectual virtue depends on the degree to which self-understanding is required in the pursuit of knowledge. To some extent it is always required. Even in the most abstract fields, more is required than recognition of one's sheer intellectual abilities, since such attributes as competitiveness, resentment, and timidity will affect the kinds of projects one takes on, the extent to which one makes use of the findings of others, and so on. A field like bioethics, so embodied and so deeply human, interdisciplinary as well as interpersonal, requires an even fuller self-understanding.

This raises the question of what communal structures encourage humility. (Some hold that only the right childhood, or compensatory psychotherapy, can produce the inner ease which is vital to it, but I believe that a variety of other things can contribute.) If humility were a matter of having been humbled, or even humiliated, the evaluation built into academic life would foster it. Someone who had first worked in a pharmaceutical company reflected on the difference she found in the university: "Here your ego is always on the line. The article gets rejected, the talk falls flat, the students are hostile—you want to die. In the corporate world we worked together, succeeded together, failed together. I didn't take things home the way I do now."

These failures, intrinsic to academic life, are exacerbated by the trend toward an "audit culture," the ubiquitous use of student and audience evaluations, and of counting and reporting devices.[36] When I questioned, in an earlier chapter, the amount of time most of us spend working, I mentioned how much of that time goes into evaluation. Let me reemphasize the point here. After I evaluate a candidate for admission to medical school, for instance, the applicant is asked to evaluate me. Did I make him comfortable? Take the right amount of time? Ask thought-provoking questions? Libby Bogdan-Lovis, our center's assistant director, spends a good deal of time simply counting: how many hours we teach; how many students; how many committees we serve on; how many public talks we give; professional offices we hold; articles, book reviews, and columns we've written; grants we've submitted; grants we've been awarded; grants we've helped implement; it goes on and on. The reporting forms come in many different shapes, count things in different ways, and serve different purposes, assuming they serve much of a purpose at all.

The idea that all this "accountability" encourages humility rests on a misunderstanding of what the real barriers to self-knowledge are. The problem is rarely a lack of accurate information about oneself. For those who *have already acquired* humility, the information can be valuable. (And sometimes it is absolutely essential, in guarding, for instance, against the abuse of students or applicants.) But evaluation does not in itself foster humility. On the contrary, it can generate a pain so great it is blinding. Someone told me of a student who had written on the evaluation form, "You are a terrible person. You should be dead." Someone else, another year, another place, read that he was a muttering old fossil. Shattered, he took early retirement. Of a guest speaker with a minor physical handicap, one student wrote, "What does a cripple know about medical ethics?" Patricia J. Williams talks more frankly about this aspect of university life than anyone I've read. An African American law professor, she received evaluations that commented cruelly on her appearance; those about her teaching were contradictory: she was too slow, too fast, too permissive, too demanding. "I marvel, in a moment of genuine bitterness, that anonymous student evaluations speculating on dimensions of my anatomy are nevertheless counted into the statistical measurement of my teaching proficiency. I am expected to woo students even as I try to fend them off; I am supposed to control them even as I am supposed to manipulate them into loving me."[37]

It's not as if written evaluations are the only way of knowing what students think. Many of us remember with chagrin our first attempts at lecturing in medical school: students reading newspapers, or chatting casually in the back of the room. Legend says some have rolled marbles down the steps toward the podium, or (my favorite) set ducklings loose. The dynamics behind these actions are complex; medical school is punishing and students in pain will strike back when they can. But it is no wonder that someone reminded me forcefully that our job takes courage, not just humility. The two are closely intertwined.

I need to emphasize that teaching medical students, especially in small groups, is usually rewarding, that student evaluations, in medical school and across campus, generally range from appreciation to (occasional) adulation, and that they can be useful.[38] Furthermore in my experience medical schools handle evaluations more sensibly than do academic departments: personal attacks are weeded out, and the emphasis is on how to improve the course (always a joint enterprise) rather than on rating the individual instructor. My point is that the ever-increasing amount of formal evaluation within the academy has little to do with encouraging humility. I should also reiterate that intellectual virtue as I understand it is needed for everything we do, not just for our research and scholarship. Intellectual virtue helps us see the world as it is, and teaching demands, besides a body of knowledge, attention to *students*. It demands trying to understand their hostilities, worries, enthusiasms; their way of viewing the world. Without humility it is difficult to see students as anything but threats to ourselves.

If intensified "accountability" does not foster humility, what will? Some examples may help. First, one from medicine. Lucien L. Leape and David Hilfiker have written powerfully about the cultural forces that interfere with the acknowledgment of error in that field.[39] With their work in mind, Tom Tomlinson has designed an innovative unit for the ethics course in MSU's College of Osteopathic Medicine. The students read Hilfiker and Leape, and then hear a panel of physicians discuss errors they have made, usually quite serious errors. The panel is always a moving experience, as mature physicians talk about mistakes of long ago that still haunt them. Nothing in the panel suggests that these mistakes are trivial, but the clear lesson is that even very good physicians will make them, and can live with the knowledge and learn from it. The panelists model what it is to recognize one's own fallibility and grow as a result.

Next a pair of examples from teaching. When Jamie first taught medical students, she apparently evoked deep hostility. When we talked, she reported that her student evaluations were the worst she'd ever gotten (she had taught two courses during graduate school). She met with the chair of her department who said of the evaluations, "Yes, these are bad. What's your side of the story?" He then advised her to observe one of the more successful teachers in the unit, not having noticed that she was already doing that. As she reported this conversation, which had deepened what was already close to despair, I suddenly recalled a time early in my own career when something similar had happened to me. But the difference was in the attitudes of her department chair and mine. Mine sat with me, read the evaluations, laughed at some and interpreted others; it was obvious that he felt we could learn from them, but was not worried about my basic ability.[40] I tried to offer Jamie that same kind of support, telling her of my own failures and of how badly they had hurt. Perhaps it helped. I have come to believe that disclosure of one's own failures, especially to those who are younger, is an important way to support people facing their own limitations.

Humility is just as important in the other activities of bioethics as it is in teaching. A doctor who can accept her own fallibility may be more open to insights from outside medicine; an academic able to recognize his own inadequacies should be more open to help from other fields. But this inner ease is only one of the intellectual virtues demanded for interdisciplinary work. It is time to talk about ways of encouraging others.

Intellectual Virtue in Interdisciplinary Practice

A highly respected administrator, now retired, told me that his practice had been "to hire people with warts; people who would speak their mind. That's so rare. The point is to hire people with different outlooks, and then manage the differences." His approach is supported by both empirical research and philosophical analyses. A large empirical study some years ago, for instance, suggested that two factors are crucial for fruitful interdisciplinary work: open discussion of disagreements, and longevity.[41] The first, arguably, is required for any intellectually virtuous community. Lorraine Code, drawing from the history of philosophy, points to the importance of "gadflies": thinkers who are irritating, who poke holes in conventional wisdom, and are therefore crucial for growth. Citing Socrates (who gave us the gadfly metaphor) and Nietzsche as examples, Code argues that a virtuous community will make space for challengers as well as conservers.[42] One implication is that we all need to be challengers, acknowledging uncertainty and even focusing on it. Such candor provides energy; it encourages us to ask new questions. Interdisciplinary work may be particularly effective here, since although it cannot always supply crucial basic information, it can highlight what is missing. On the other hand, instead of revealing deficits, interdisciplinary work can sometimes camouflage them, through differences in discourse and misguided forms of courtesy.

The second factor in success identified by the study was longevity, which would correlate closely with the virtue of perseverance. The finding confirms common sense: working across boundaries becomes easier with time, and mutual understanding deepens, as each party gets to know the assumptions and conceptual tools of the others. The more basic question, however, is what makes longevity possible. This desideratum pushes in a different direction from the first, since (to mix metaphors) too many warts on too many gadflies will dissolve the community. Only a deep courtesy and respect will carry us through. William James, who himself crossed disciplinary boundaries, phrased the need eloquently: "No one of us ought to issue vetoes to the other, nor should we bandy words of abuse. We ought, on the contrary, delicately and profoundly to respect one another's mental freedom; then only shall we bring about the intellectual republic."[43]

What is this mutual respect like? How can it be encouraged? Sometimes disrespect is overt and acknowledged (some philosophers among themselves habitually insult, say, doctors or social scientists or economists; similar things must happen within other disciplines). Disrespect is more deadly, however, when it is unrecognized. There is an old joke about a neurosurgeon and a historian: after introductions at a party, the doctor says admiringly, "I've always been interested in history. I think I'll take it up after I retire." To which the historian responds, "And after I retire, I plan to take up neurosurgery." I've heard similar unconsciously insulting remarks made in bioethics; I'm afraid that I've done it myself. "I'm doing something like anthropology," I once said, describing this book project. The resulting dead silence signaled that I had gotten something very wrong. In contrast, I later heard Ed Pellegrino get the same sort of issue exactly right. Accepting a lifetime-achievement award, he reminisced about the field's early days, but then emphasized the difference between his memories and history: "The role of an elder is to remind you of early days that may be forgotten. But this isn't history: it's not history until subjected to a historian's analytical intelligence, and studied by an outsider."[44]

Another manifestation of respect is one I've already mentioned: recognition that one cannot evaluate work done in other fields, and a consequent caution about making use of their findings. Scientists learn not to quote data unless they have themselves evaluated the studies; we are in a harder position when we do not even know *how* to evaluate sources, research design, and so on. It has helped me to learn that the "editorials" in medical journals are expert evaluations of studies reported in the issue: they will discuss a study's limitations, the way it reinforces or contradicts other findings in the field, and where, all things considered, the weight of the evidence now points. It has also helped to be in contact with people working with evidence-based medicine, all the more so since my sources keep a critical eye on EBM itself.

Since outsiders are often unable to evaluate a discipline's work, it is important for each to be self-critical. I saw a fine example of that kind of self-evaluation at a conference, a small working group sponsored by the Hastings Center and concerning the patenting of genes. During the conference, theologian Ronald Cole-Turner roundly criticized clergy from his own tradition for misunderstanding and misusing it. (They had issued a public statement denouncing the patenting.)[45] No one from outside that faith community could have done it: the rest of us lacked the tools and the standing. Without the critical interface of Cole-Turner, the only options open to the rest of us would have been dismissal of the logic or respect for the emotion in the public statement. We could not have critically engaged the ideas on their own terms.

But criticism will not always come from within. Criticism from other parties and other fields poses special challenges. It is likely, for all the reasons I've just given, that the critics will either misunderstand or sound like they misunderstand the activity they are criticizing. What can make this situation better? To begin with, those offering the criticism, keeping in mind the distance it must cross, can choose their words accordingly. In the space of a few days I once heard two anthropologists criticize bioethics projects. One said, "As an anthropologist I could see they were being ethnocentric." The second, speaking of a different project a few days later, said, "I've been trained to ask whose interests are being represented, who gets a voice." These points are not exactly the same. But the second seemed to me more useful and more respectful, in part because it was more specific. My instinctive response to the first was that I would never make the corresponding remark, "As a philosopher I could see they were conceptually muddled." In fairness, I should add that the first speaker simply has a cheerful, blunt manner, and that she went on to some specifics: "They weren't dealing with what mattered most to the patients." But even this seemed to me to beg some questions, since she was assuming the appropriateness of a certain goal.

On the other hand the problem could have lain with me, the listener. The self-understanding I discussed earlier under the rubric of humility is crucial. One needs to remember how common defensiveness is, as well as the problems our not-quite-shared language can raise. It helps to use what one of my colleagues calls "pro hominem" reading: if *ad hominem* is an attack on an argument based on defects in the person who made it, then *pro hominem* is a charitable reading of an argument based on one's respect for the person who made it. At times "they" will seem to "us" frustrated into incoherence or cruelty, and it will be up to "us" to decode the incoherence. Interpret, translate, reconstrue: try to find the truth, or at least the intent, in what was said. Trying to understand the writer's purpose can be a wonderful tool for this kind of reading, as I discovered when investigating theories of moral development.[46] What sounded like the silliest form of relativism (counting the search for a right answer as a sign of immaturity) was part of a theory with the gravest sort of intent: to honor intellectual growth that moved from blind acceptance of authority to responsible questioning.

Sometimes some psychological interpretations will also help. Lainie Friedman Ross, a pediatrician who is also a philosopher, once called the tensions between medicine and philosophy "stranger anxiety," a stage in infant development. Stranger anxiety can be cured by acquaintance, and more permanently by maturity. A less charitable, but often true, interpretation sees tensions between disciplines as turf battles. But the deepest description of this problem came to me, as one would expect, from an anthropologist. She was describing differences between herself and sociologists, between herself and doctors, and between both of those groups and ethicists. "I make broad statements based on a small sample; it drives sociologists crazy, but I operate on certain assumptions about the relationship between the individual and culture. During one consult an incompetent patient's wife was passive while his mother bossed everyone around. I wanted to know who, according to that subculture, had status and authority. In some families the wife isn't even really a relative. But the discussion was all psychological: how do we give the wife a voice? I was the only anthropologist on the case and they just couldn't hear me. We'd have to work together a lot longer before I could seem like more than a black box to them. Our perspective is so different: docs want good quantitative research, and I just don't do that." She went on to comment that the pursuit of knowledge, for doctors and for sociologists, is the pursuit of something new; the frontiers of knowledge have boundaries that can be expanded. So ethics and anthropology, for whom the spatial metaphor would be depth rather than expanse, won't count.

All of this, however, is the negative side of something much more positive. The anthropologist whom I just quoted did sometimes manage to communicate with clinicians, and then they found her work uniquely enlightening: they saw how much more information a narrative contains than a statistic does. Whether presenting a different perspective or trying to understand one, more is required than courtesy. We need to welcome different modes of inquiry, to realize the joy they can provide. I first understood this possibility in teaching a seminar for graduate students about interdisciplinary work. The assigned readings included literature, anthropology, sociology, philosophy, and history. Wanting to provide a unified approach to all these different fields, we had drawn up a template for analysis: What is the author trying to accomplish? What tools does he or she use? With what success? The framework was logically impeccable, but too abstract to be really useful. I realized vaguely (because my undergraduate major was in English) that in studying literature students need some help noticing characterization, point of view, figurative language, and so on; and I invited a guest speaker to address these topics. Since that presentation worked well, I invited someone from history to help students learn to read in her field. And that class was a revelation to me. I asked Jody Ross, a doctoral candidate and gifted teacher, to talk about chapters 3 and 4 of the assigned text (Tom Cole's *The Journey of Life*).[47] She found my request impossible. As a historian, she had to, first, read the introduction and the conclusion to the book as a whole, and secondly *scrutinize the footnotes and references.* The first made sense to me; it's a useful approach to most difficult reading. But read the footnotes? Before, or even without, reading the chapters? Jody kept saying, "I love his sources!" Her enthusiasm was perplexing but contagious, and soon I understood; a historian's primary tools are his, or her, sources. The nature and success of the work depends on the adequacy of those sources to the claims made. It seems so obvious, once written out. But I come from a field that attends very closely to the logic within the text, and only to the text. Jody's class lit up my mind. Interdisciplinary work creates the possibility of such epiphanies. My discovery that day—that in our various fields we seek different goals through different means—complements my earlier discussion of the "languages" of bioethics: what seem to be, in Wittgenstein's words, common language-games, can be quite different. We do not always understand one another, but we can always get closer to doing so. Recognizing the distance is the first step to crossing it.

For all of these purposes—avoiding disrespect and defensiveness, welcoming other kinds of work while exercising care in drawing from it—the most important means is to engage with one another on the practical level. Again this is a call to *praxis.* Concrete problems (a case, or policy, or custom) belong to no one field. They allow us to focus on something outside ourselves, to avoid grand theoretical debates, and instead to forge useful tools for particular purposes.

An interdisciplinary community, then, in order to flourish, requires openness about disagreements, and attitudes that support longevity: mutual respect, self-reflection, and an appreciation of the rewards that lie in store. And once again, a virtuous interdisciplinary community cares for its young. Such care includes preparing graduate students for the work they will have to do to make a living. Someone recently out of graduate school told me how hard she found her first job: "I was overwhelmed. I had never been trained to write grants, use statistics. In grad school I had to learn medical terminology but not this stuff that now I have to use every day." She worked as a research associate in a project aimed at improving care for the dying; the project was thoroughly interdisciplinary, asking descriptive questions about what dying is like, evaluative questions about what would count as better dying, and practical questions about how to change things. The requirements to learn medical terminology in her graduate program, she thought, stemmed from the days when we shaped our jobs to the needs of individual doctors, in ethics consultation. The work these days is larger, and she had not been prepared for it.

Exactly what constitutes an adequate training is a vexed question. I once heard someone with two professional degrees, one in philosophy, make a complete hash of a basic point in ethical theory in a public presentation to clinicians. I was taken aback and puzzled, and finally decided that the problem lay in certain facts about his training. Philosophy was his second field of study, and he had always made a living in his first. Very much of what one learns in any field comes from practicing it; he had never "practiced" philosophy—never taught it, never published in it. (Teaching especially is invaluable, since it demands so much distillation and clarity. Undergraduate students cannot fill in the gaps in what one says, as peers will.) He taught and published regularly in bioethics, but, as I've argued throughout this book, that is not the same as philosophy. After hearing his presentation I once again realized that learning any field—even so textual a one as philosophy—is more than a matter of studying words on paper. It demands activity and interaction as well. How to provide such experience in several disciplines for the next generation is not an easy question to answer.

Intellectual virtues, then, are acquired dispositions, in individuals and in communities, that promote growth in knowledge and understanding. In bioethics and the related humanities, the objects of our inquiry are the moral dimensions of the world: what is of value, how it is or is not respected, and what practical measures would improve that picture. Our tools of understanding are multiple, and there is no algorithm for the way they can be used together. The virtues we need in our enterprise include courage, honesty, humility, perseverance, and mutual respect. The rewards of these virtues are great: they will help us accomplish our goals and provide us with intellectual satisfaction as well.

11 A New Millennium

I end this book in a new millennium. Of course the time that has elapsed since 1994 is really less than a decade, but the years have been significant. Some of my perceptions have changed: I am less in awe of the circles in which I move, less impressed by the fact that sometimes I write things that end up on hospital charts. I have spoken to journalism classes and journalists, written about bioethics' interaction with the media, and become a little uncomfortable at the forms my earlier criticism took.[1] More than my perceptions have changed, however; my life has, and so has the world around me.

Personal Horizons: My Brother's Death

My brother, Pat, died about a year ago of the cancer diagnosed shortly after I began my interviews for this book in 1996. His struggle instructed me in the complexity of what it means to respect patients. I found it painful and ironic that things for which I have advocated professionally he did not particularly want: sufficient morphine, candor from his physicians, comprehensive primary care, a demythologized view of early clinical trials. He wanted to live, and nothing else really mattered. There was one point of convergence, however, between my professional concerns and his personal ones: at every stage of the disease it was clear that malfunctioning health care systems contributed to his suffering. He and his wife, Liz, constantly faced a medical system in disarray, although (because?) they were treated at nationally prominent medical centers. Lost records. Information not available when they wanted it, or thrust upon them when they had asked for a few days' reprieve (the doctor who blurted it out being three people removed from the one whom they had asked for time). When simply getting to the hospital cost an hour's worth of pain, finding that their appointment had been rescheduled. Forced to lie waiting on a cold gurney without adequate cover, even when the cancer had spread to his bones. Having to call a nurse on the other side of the continent to find out whether his wracking chills meant he should go to an emergency room; then eight hours waiting to be examined, shaking uncontrollably, only to be seen first by a nurse who wanted all his clothes off *now.*

A fourth-year medical student wrote me recently of her awakening to this kind of thing: "Now that I'm 'in the trenches,' I see the need for shifting focus to what the patient needs. The medical system (yes, albeit peopled by wonderful caring people, but individuals do not equal The System) is so *not* about the patient. A patient submits his or her body to The System and The System does its thing with it. Often very well, but that there is a person and a life attached to the body is really not a consideration. Hospitals are amazing places. Temples of The System." These problems seem to be nobody's problems, except the patient, his family, and the occasional unreconciled clinician. If a patient complains, it's a concern for Risk Management or Patient Satisfaction, not for the ethics people. But this suffering is as real as pain, with which ethics committees deal regularly.

My brother's struggle led him deep into a spiritual quest. While he was embracing Byzantine Rite Catholicism, my own interest in Buddhism deepened, not as a theology, but as a practice and a set of perspectives. These strengthened my conviction that moral development is more holistic and embodied than either ethics or philosophy has understood, a conviction that has shaped this book.

Professional Vistas: The Third Annual Meeting of ASBH

Within bioethics and the related humanities, these years have been exciting, even exhilarating. The emergence of ASBH unified the field and gave it critical mass. Sessions at the annual meetings now are of higher quality, and each year the crowd seems more filled with energy than the year before. The focus on end-of-life issues, which I criticize in chapter 7, has lessened, and the muddier methodological discussions ("principlism," narrative, and so on) have clarified or abated. The sense of life and growth is in marked contrast to the American Philosophical Association (APA), an older and quieter organization. Unfortunately another contrast between the two remains as well: Bioethics remains, for me, a somewhat tenuous intellectual endeavor. The book exhibit (where publishers display their wares) is still much smaller than at the APA, and it vanishes by Saturday. (As I approach the end of this manuscript and look at everything I have pushed aside, I realize that bioethics journals are piled high. It is not yet material that I find compelling.)

In her presidential address at the ASBH meeting in October 2000, Laurie Zoloth called for the organization to take an official stance on certain key issues, especially on the question of national health care: "Make it a liturgical device," she said, meaning a topic we each bring up in every talk we give. I had been doing something like that for years, and suddenly felt I had found an ally. Her call was met by a standing ovation, if also by criticism. Thanks to initiatives begun during her presidency, ASBH will soon articulate its public role more self-consciously. I believe it will find a middle way between activism and "informing the debate," a way more tentative and reflective than a political group can be, but more engaged than a purely scholarly group can be. I believe, in other words, that the organization will find a way to participate in a kind of national moral development about health care, health science, and health policy.

Other forces as well feed the bustle and bloom at ASBH. There are jobs, and job mobility, as there are not in the other humanities. There are new questions in the older fields as well as in bioethics, but ours are more practical and carry a sense of urgency. Some of our new issues involve research ethics; others arise from increasing attention to the internal life of the field, to the kinds of questions I have addressed in this book. Sue Rubin recently led an ASBH session called "Pulling Back the Curtain," in which panelists talked about what motivated their own work, and speculated about that of others; presenters encouraged attention to the subterranean and personal factors that shape the field. There has been an influx of money into bioethics as corporations (especially in biotechnology) have begun soliciting and supporting ethics projects. Companies give money to centers, commission research, and hire people for IRBs, and all this involvement raises questions about conflicts of interest. The issues are urgent and energizing.

This vigor does not change the fact that bioethics, like any profession, is in a sense just one large pecking order. During my first day or two at any ASBH meeting, I am painfully sensitive to the need to please and to achieve, within myself and in others. I meet Jack (who rejected my manuscript!) and Jill (who ignores my friend's work) and John (who worries that spending time with his family keeps him from networking). One year Dan Brock gave a plenary address focusing on the importance of socioeconomic status to health, and I was torn between feeling scooped and being glad this essential point was getting such high-profile attention. Our graduate students last year commented on how "clubby" the atmosphere is, on the ubiquity of what seemed to them like cliques, although they also commented on the lack of snobbishness. Sometimes I'm afraid that my, our, very purpose at the annual meeting is simply self-promotion, or more charitably getting good people (that is, the ones close to us) the status they deserve. One year, as I was experiencing my yearly shock at all this fervor, I happened to hear a paper on Freud's narcissism. It helped me sort some things out: a healthy self, he said, is invested in others and engaged in the world; narcissists, in contrast, are invested in *images* of themselves. A professional conference encourages both attitudes, and encourages confusing them.

I suppose it's not just conferences but professional life itself that encourages the confusion. A friend told me over coffee that a senior colleague had warned her, "Your career's off track. You've got to focus." My friend believes the older man was projecting his own sense of inadequacy, but she was clearly still bothered. Just how should she focus, and why? For the moment she is limiting herself to being aware of why she makes certain choices, for example, to speak to a high school group. Giving the talk will do nothing for her résumé, but it's good for the students, and it's not the kind of thing her untenured junior colleague should be doing yet. For him it would be a dangerous loss of time. My friend is invested in the world, and helping her younger colleague invest, as he must, in images of himself.

As we move uncertainly toward professionalization, we move both toward and away from community. The ASBH sessions in October 2001, as they had in the preceding annual meetings, included far more panels than papers. While papers were submitted as full-length drafts, and so were judged on full evidence, panels were submitted with brief descriptions and the names of the panelists. Rumor said the program committee chose panels largely by the stature of the panelists. "Big names" are believed to guarantee a "big draw" —and perhaps also interesting presentations? The results were predictable: a predominance on the program of people who were already known. On the other hand, ASBH also makes real efforts to help graduate students, including formal mentoring sessions and a task force on graduate programs.

The problem of overwork remains. A friend tells me she and her husband (in a different profession) both work seventy-hour weeks, "but we're fully engaged in what we're doing. We like it." A newspaper (which I've now lost) quotes an anthropologist to the effect that we're in the midst of an unnoticed transformation. "What does it mean when a three-year-old says to her playmate, 'Don't bother me now, I'm working'? We're making work the be-all and end-all." As I meet our freshly minted Ph.D.'s in their first jobs, I am often struck by their exhaustion. One in particular, slender when I last saw her and now thin, told me that she does not expect the exhaustion ever to end. She likes her work, and she's in good spirits, but it's painful to look at her. I expect to retire sometime in the next decade, and I wonder if once this manuscript is done I'll be able to find more balance in my own life. I take heart from someone I met at the conference, a friend in his fifties leaving the conference midstream. "I come, I schmooze; then I leave. I want time with my family." He's at peace with this.

I'm not always at peace with my own priorities. In writing this chapter I looked back at my notes from earlier years, and realized that I had seesawed between balance and exhaustion. My notes for late October 1999 (the month of the ASBH conference) include an admiring but appalled description of a colleague who had risen at 4:00 A.M. to work on a grant proposal; by the time we met, her anxiety was powerful and contagious. I had myself been badly overextended for the month, a month that culminated in my giving a disastrously bad, both over- and underprepared, presentation off-campus. My notes mention pain in my legs, my shoulders, my head; a stupefied exhaustion, a need to sleep before I could even pack for the conference. The following year was different. I said "no" repeatedly, and cleared some periods of sustained working time, almost adequate for the work I needed to do. But the year after that I was back again to an impossibly overcommitted semester. When I told my nieces and nephews I would have to miss Halloween once again, their mother said, "It's been two years since you were here!" She meant two months, but it seemed like forever.

These personal and professional problems paled, of course, on September 11, 2001.

National Horizons: Terrorism and the American Response

On the day of the attacks on New York and Washington I was, like everyone else, almost paralyzed by what I saw and heard. I nevertheless understood one thing: the perpetrators lived in a world of absolutes. They had created a picture of others—us—as irredeemably evil, and of themselves in contrast as righteous. During the day of September 11 I heard world leaders keep that polarity, simply reversing identities: We were the good, and "they" the evil. I do believe the attacks were evil, and I have no interest in justifying them. But the abomination of what was done could not insure that the response we chose would be justified. If anything the extremity of the situation made it likely that we would fall victim to our own oversimplifications and choose badly.

I was haunted by memories of too many simplistic, dismissive, and even angry conversations. As it happens, the people I love span a wide spectrum of political, religious, and ethnic identities: Byzantine Catholic, Shiite Muslim, atheist; right wing and left wing; Jew and Arab. I remembered an Iraqi woman widowed by the Gulf War chanting rage to her grandchildren; a Christian pastor announcing that Muslims believe they will go to heaven if they kill a Christian; a cousin who suggested that the U.S. government had bombed the federal building in Oklahoma City; left-wing friends who believe that everyone on the right is stupid or pathological. I remembered not only these proclamations but also my own silence in response. I took a small private vow to speak up.

Within a week I realized that my personal and civic concerns were part and parcel of my professional ones. The concept of moral development that I present in chapter 6 includes the idea of cognitive complexity, developed by theorists interested in what went wrong in Nazi Germany. Moral development was stunted there in part because of authoritarianism, a monstrous overdevelopment of the general need to abide by the rules. Part of what goes wrong in contemporary terrorism is an equally monstrous overdevelopment of other human needs: for clarity and for commitment. When Newton formulated the laws of motion—when he captured thousands of observed facts in three simple principles—he transformed our view of the physical world in a deeply satisfying way. The moral world does not admit of such a reduction, but most of us would be comforted if it did, and in that fact lies danger.

On September 11, 2001, I felt that the United States had been sucked into a river of hatred and blood whose current we could never escape. I knew that our foreign policy was part of the context for the attacks. I also knew we had to respond to them, but could imagine no way of doing so that did not simply feed the hatred from which I wanted us to escape. I ended the week in despair.

But the despair lifted. Over and over again, in my own life and across the nation, I heard evidence of the cognitive complexity I find so important: not what the Queen in *Alice's Adventures in Wonderland* achieves (believing impossible things for half an hour a day) but a recognition that understanding is always limited and perspectival, and at the same time always capable of deepening.

Not surprisingly, in the immediate aftermath of the attacks I did hear oversimplifications, and even pure fabrications. E-mail circulated within a Jewish community claiming that patrons of a local Middle Eastern restaurant cheered as they watched the World Trade

Center towers collapse. I heard of Arabs who believed that Israel was behind the attacks, and that 4,000 Jews had been warned not to go to work that day. Some on the political left attributed responsibility for the attacks to U.S. foreign policy; some on the right found any criticism of U.S. foreign policy treasonous.

Almost immediately, however, I heard other voices, not only intellectually responsible but also compassionate, and I soon realized that the problem was not just oversimplification but an accompanying hardness of heart. I described my fear of an inescapable torrent of hatred to a friend who replied immediately, "There's a far wider river of compassion." She was right. There's nothing sentimental about her point; we would not have survived as a species if it wasn't true. The evidence was everywhere, from my student late for class because he was giving blood, to the memorials on campus that Friday, to the national outpouring of help for victims on the East Coast. Within hours of the attacks e-mail went out: we must protect our American Muslim neighbors. From the beginning political leaders distinguished between the Islamic mainstream and the fanatic hatred of particular men. The Jewish community I mentioned called itself to task: Arab Americans in that restaurant of course had *not* rejoiced, and some of those who received the e-mail rumor made sure the truth was published (literally—a newspaper story covered both the rumor and its halting). The grandson of the Gulf War widow I mentioned earlier flooded me with stories of mercy and love from the Islamic canon, and especially of compassion toward one's oppressors. Nearby, his grandmother listened, with appreciation.

People became able to say that although this was not Islam versus Christianity, or Islam versus the West, religion was part of the story. We were dealing with religious, not just political, extremism, and religion deepens the intensity of whatever it touches. As for the role and responsibility of the West, it became common knowledge that our Middle Eastern policy has often been disastrous, morally as well as practically, yet most people understood that what happened on September 11 had other roots as well: a crisis of modernity in the Middle East; political expediency among leaders there; individual psychopathology (in, for instance, Osama bin Laden).

Dealing with these tangled roots while in no way absolving the perpetrators of full responsibility was a complex task, but one many Americans seemed prepared to accomplish.

Through late September and October, I paid attention above all to the way the Bush administration approached its terrible new responsibility: to respond, to take action, to bring about death. I did not appreciate the metaphor of war, but it stuck. In spite of the inappropriate language, however, the administration took surprisingly nuanced and careful steps in preparation for military action: it searched for alliances, constantly distinguished Muslims in general from the terrorists in particular, differentiated between the people of Afghanistan and the terrorist network, provided humanitarian aid. The administration understood that more was at stake than eliminating al-Qaeda; that what came next in Afghanistan would have implications for our children's children. I tried mightily not to give credit for any of this to George W. Bush until someone reminded me that his was the final word, and that he chose from the competing suggestions of advisers he had himself chosen. One of my own simplifications of the world had gone, and I was glad for the unexpected opening of my heart.

What eventually happens in Afghanistan and in the Middle East, and how much credit or blame will ultimately accrue to the United States, no one can now say. But I was reassured by what I heard during the fall of 2001. In comparison with, say, 1941 or 1968, we have matured.

Teaching

The national discussion of abortion, however, remains angry and ugly. Opponents simplify the issues and caricature one another. Our medical school ethics courses now address abortion, trying to complicate the way our students understand it. (Our courses in both medical colleges have broadened significantly in other ways as well; the narrow focus on end-of-life issues that I criticize in chapter 7 is gone.) The "reproductive ethics" unit focuses not on the rightness or wrongness of abortion, but on challenges to integrity: doctors on any side of the issue face hard choices. We added the unit with trepidation, wondering if the classrooms would explode in anger. To the contrary, we found no throwing of slogans, no insults, no feeling of being silenced; I happened to have in my group the president of Medical Students for Choice as well as the very conservative class president—each highly articulate, principled, and respectful of the other.

Why did we do it? To call attention to an area where things are changing, quickly—as hospital systems merge, and one partner is religious (especially Catholic) while another is not, decisions must be made about a whole array of reproductive services. They may effectively disappear over large geographic regions.[2] Or, if the opposite happens, the religiously affiliated partner may feel complicit and contaminated. Fewer and fewer doctors are being trained in abortion procedures.[3] Designing the course as we did, we were not only promoting moral reasoning but directing attention, that is, encouraging moral perception. Beyond that, the teaching unit asks people (from all points on the spectrum) to think about particular policy and political stances, so we are encouraging action, albeit not in a particular direction. We are supporting moral development as I describe it earlier in this book. And just as I described it there, the development is mutual and reciprocal: I came to realize that I had avoided teaching the issue partly because it included questions I didn't want to think about myself. Having faced them, I have brought up the issue in my course for undergraduates and even in conversation. I have grown with the students, in understanding, and in taking action: that is, raising an issue that is often avoided because it is difficult to talk about well. I realize that the more usual model of our role, that of teacher, would also accommodate this action—but that is because, as I've argued earlier and elsewhere, the role of teacher involves more than helping students learn moral reasoning.[4] Teaching practical ethics has always been a sort of shared engagement in moral development, heavily weighted toward the moral reasoning component, but not confined to it.

In other teaching encounters, I fared less well. (For me, teaching and public speaking are like ice skating: there are wonderful stretches of grace and balance, punctuated by hard falls.) Speaking to a group of residents on ethical issues in the use of placebos, feeling comfortably prepared because I was using work by Howard Brody, which I have long admired, I was left speechless by the first question: "Is there a metanalysis of all this?"[5] I took another hard fall in working with veterinary students when I discovered that a question of clarification ("Are you taking the welfare of the animal into account?") had been heard as accusation: "You are not taking the welfare of animals into account." The two experiences led me to expand the "pidgin" metaphor a bit. My language was foreign to these audiences, partly because we had no trading relationship yet. My first job was to convince them I had something they wanted. To convince them of that, I would have to speak *their* language, because what I have to offer can be shown only in words. After that we could—maybe—formulate a language with which to negotiate an exchange.

In another closing of a circle, I have returned to teaching undergraduate bioethics, as I was doing in the fall of 1994. My developing understanding of the practice of bioethics has led me to make major changes in the course. Minimally (and not really a change), I now want the students to learn not only to recognize and reason about ethical dilemmas, but also to recognize that an ethical consensus exists (where it does). In addition, because most of them will become doctors or nurses, they need to know how hospital policy is formed, what its limitations are, and where to find it. They need examples of all that from local hospitals. I want students to learn, to think more clearly about not only allocating scarce organs, but also what the United Network for Organ Sharing is and how it works. They should know not only what democratic deliberation is, but the ways in which it can be subverted by tone of voice, seating arrangements, and choice of rhetoric. They should know what the Office for Human Research Protections (OHRP) is, what its regulations are—down to seeing the fine print itself—and understand what the IRB process is like for both researcher and board members. They should consider the implications of OHRP having such a small staff. They should see a good informed consent form and a poor one. I am not able to do all of this, of course, and even doing some can be at the expense of philosophical analysis. If I have to choose between talking about OHRP's small staff or tackling the questions that arise in applying U.S. research standards overseas, I do the latter. But skill in recognizing, criticizing, and constructing moral arguments is not all that students need. They need to understand bioethics as a practice.

Organizational Ethics

That bioethics is a practice shows particularly in a new facet of the field, called organizational ethics. Addressing questions beyond those about the care of individual patients, organizational ethics looks at the moral dimensions of an institution's corporate decisions, culture, and structure. Whether a hospital should advertise, how it should deal with unions, who should have a voice in corporate decisions, how competitors should be treated, what environmental responsibility means in health care—this set of questions is new to bioethics since 1994. I've watched the initiative with some interest. The entrepreneurial ("gold rush") aspect of bioethics came first and has been useful, as people hurried to convene conferences and publish books. At first I was startled and somewhat cynical about what was happening; some of the people plunging forward knew as little as I about organizations, and the results were often naive or irrelevant. By now, however, I recognize the value of their boldness. There are some good books from a handful of people with the background to know what they are talking about. But there is also a body of published cases. I have come to think there is nothing more important, at the beginning, than a good set of cases. From these the analysis can begin, the fledgling organizational ethics committees can begin to educate themselves, a literature and counterliterature will develop, and perhaps in ten years there will be a sufficiently coherent set of activities that organizational ethics can be counted as an aspect of the practice of bioethics.

My own part in this development has been complex. I was part of a multihospital project to hear, and give an analysis of, the moral distress in which most nurses today work, situations I describe briefly in chapter 7. Our task force worked for years on the issue, and constantly found ourselves walking into firestorms and firewalls, as we badly misread institutional realities. A report, which I thought so important that I took months away from this manuscript to do it, remains embargoed for reasons of confidentiality (really, I think, from exaggerated fears on the part of administrators). The report registers something close to despair among nurses, who are stretched beyond endurance every day and given no voice to express it. I have watched seasoned nurses leave the field, finally abandoning hope. In Flint, fifty miles away, a nurses' strike lasted for weeks; in Lansing one was threatened. The nurses I know across the country, aware of how understaffed hospitals have become, are making sure that a family member is at the bedside of any hospitalized relative twenty-four hours a day. Yet one of my colleagues, sitting on a statewide commission to improve safety in hospitals, found the commission unaware of the impact of nurse-staffing ratios on patient safety. (It did have data about physician staffing, of course.)

The project of which I was part felt like a failure to me. If organizational ethics had become a mature field by the time we started, we could have drawn from it: we would have had the leverage of a national consensus about the legitimacy of initiatives like ours; model structures of task forces, or organizational ethics committees, from which to draw; a history of the successes and failures of others from which to learn; and an ethical consensus about, say, the process of drawing up budgets, about who should be a party to it, and what kinds of questions should be on the table. In the case of nurse-staffing ratios, all this hypothetical effort still might not have been enough, since the economic realities are so dire. (Nursing salaries are the largest single item in a hospital budget.) In any case the country now faces, for the indefinite future, a severe shortage of nurses. But even a partial success, in the form of more open institutional discussions, would have been gratifying.

The story is not finished, however. Working on the report with clumsy tenacity, we eventually managed to be heard by one of the hospitals on one point: organizational ethics issues exist, and it would be good to have a way to talk about them. As a result, organizational ethics will soon be officially recognized at that hospital. Those of us who worked so long on the project have been strengthened by one another, grown morally together. We learned together to see what needed to be done, and drew from each other the stamina and skills to do it.

The same project helped me articulate the distinction between activism and an engaged bioethics. At one point, thinking that we should intervene on one side of a dispute, we suddenly recognized that our job was to be a resource and a challenge to both sides. We were not neutral about what should be done, but our purpose was to help others *recognize* it as well as carry it out, and simply adding muscle to one side would not have helped. In other situations, however, taking sides might be appropriate, since we would never choose a side without making our reasoning clear, and never close off further reflection about our choice.

All over the country, efforts like these are proceeding, and eventually a practice will emerge. But it will not come easily. The activities of getting people to pay attention to the enterprise of ethics, to take it into account as they set up policies and procedures, and to join in ethical discussion are entirely different from analyzing what decision should be reached. That these things have been accomplished in patient care and in research is the result not just of individual and group effort but also of specific historical events: court cases, scandals, and so on. There are a few clues, but only a few, about what might help organizational ethics develop. Institutions seem to accept such activities in certain kinds of circumstances: when a senior administrator, on his own, has an interest in it; when a "sentinel event" (a threat to the hospital's image or pocketbook) is seen as a matter of organizational ethics; and when the initiative somehow draws energy from "compliance" efforts (the need within hospitals to comply with many different kinds of regulations). As for the form the practice of organizational ethics will take, at the moment people are using the committee and consultation model that has worked in clinical ethics; it's a useful place to start.

I have seen the moral impact of organizational structures in other ways since 1994. Then I wrote about "giving a voice" to women who were resisting medicalized childbirth. I do continue to express that perspective, but I no longer think that giving a voice to people means speaking for them. In 1994 Libby was my student. Today she is a colleague, and speaks most effectively for herself. There are parallel stories elsewhere. One center tried to help an immigrant group be heard by local doctors and nurses, and was unsure its project had any impact. "But what did help is that members of the community are going to medical school. Some have graduated; they're making a difference." It's well and good to hope to provide a voice for someone else, but the effort can be self-deceptive and condescending as well. Here, as so often in bioethics, the deepest work is concrete: helping people find a legitimate place at the table, rather than trying to speak on their behalf. This can be enormously hard; it can be impossible. In Libby's case what made the difference is that MSU has an employee category called "academic specialist," a kind of midway slot between faculty and staff. "Specialists" often have master's degrees and perform a variety of jobs, from running labs to doing academic advising to, in Libby's case, helping to administer the unit. The job description is sufficiently open that someone with her initiative and ability can teach and do research as well.

In contrast, the architecture and geography of MSU are obstructions to some of our work. The vast campus makes casual interaction unlikely; furthermore, individual buildings like the one in which I work, originally built as a dormitory and designed to keep boys and girls apart, now keep faculty apart. Organizational structure, including the design of campuses and buildings, makes a moral difference. Working to change them could be a moral project.

Conclusions, for Now

As the new millennium gets under way, there are things to celebrate. The field of bioethics is prospering, with new work on issues in international research and on financial conflicts of interest. But old problems for the field remain, and new ones have arisen. Judging from editorial choices on CNN and in the *New York Times* on the web, each of which now gives daily space to "health," the American fascination with medicine and fitness has increased. Poverty and inequality, major contributors to poor health, are almost never mentioned. On the other hand, coverage of the Human Genome Project seems to have moved beyond the genetic determinism of early "Gene of the Week" stories to something more sophisticated. Hospitals are foundering, in part because the Balanced Budget Act of 1997 cut their incomes so drastically, and the current administration continues to find tax cuts intrinsically good in all circumstances. The public outcry against managed care may end some of its sillier micro-management, but has not helped educate people to the fact that health-care resources are finite. Clinicians still seem to believe that they must torture their dying patients unless the family allows them to stop. "Nonprofit" hospitals compete in wasteful and self-serving ways. Nursing remains invisible, except in its absence, and then the problems may be blamed on incompetent individuals rather than on the decisions that left the units understaffed: locally shortsighted budgetary decisions, and the national pattern of undervaluing nurses that has made the field unattractive to young women and men.

On a more personal level, Jamie—the young academic whom I've mentioned in the last few chapters—now has tenure. I, at quite a different stage of my career, end this manuscript as I began it, cherishing the daily gratifications my job offers, the sense of having made a difference, large or small. I am profoundly grateful not to share the attitude of G. E. Moore, who once said, "I do not think that the world or the sciences would ever have suggested to me any philosophic problems. What has suggested philosophic problems to me is the things which other philosophers have said about the world or the sciences."[6] The world speaks to me directly, stimulates reflections like those in this book, and rewards me. I'm told that someone carries a column of mine in her pocket. Another university asked to post an article I wrote on its high school bioethics website. A student who did not understand, now does. A doctor embraced her patient's wife, reconciled after a tearful ethics consultation. A student thanks me because she has learned to listen to those with whom she disagrees. Each of these small things attests that someone, here and now, finds something I did meaningful; if, as I believe, bioethics is about making a difference, in the end those differences must be in individual lives, even if they come about only through corporate and cultural change. Work on the large issues is hard and slow. I am, as I believe we all are, sustained in it by such everyday fruits as these.

Notes

CHAPTER ONE

1. See further comments about my choice of this term in the Preface.

2. The symposium at which the papers were presented would be the twentieth anniversary of the first Trans-Disciplinary Symposium on Philosophy and Medicine; the volume in which the essays were published was number 50 in the Kluwer Philosophy and Medicine series. Carson and Burns, *Philosophy of Medicine.*

3. Henderson, "Epistemic Competence."

4. The panel was held during the March 1995 meeting of the Association for Practical and Professional Ethics in Washington, D.C.

5. Lukes, *Power.*

6. Reiser, "Ethical Life of Health Care Organizations."

7. Hart, *Sociology of Health and Medicine.* In later years we have turned to a textbook written for the British Open University system, which makes the same point throughout. See Webster, *Caring for Health.*

8. Hilfiker, *Not All of Us Are Saints.*

9. Since then, there has been some sustained work done in establishing standards for ethics consults. See, for instance, Society for Health and Human Values, *Core Competencies.*

CHAPTER TWO

1. MacIntyre, *After Virtue,* 181–203.

2. I first heard Stephen Toulmin make this point at the Galveston gathering in 1990.

3. Toulmin, "How Medicine Saved Ethics."

4. Ibid.; Richardson, "Specifying Norms."

5. Society for Health and Human Values, *Core Competencies.*

6. Andre, "Goals of Ethics Consultation."

7. Keynote Presentation, Summer Intensive Workshop, Medical Ethics Resource Network of Michigan, Michigan State University, June 29, 1997.

8. The personal part of this story is that at Michigan State even the traditional teaching that I was assigned was different from what I was used to,

although not uncommon in research universities. The teaching format upon which I had cut my teeth—sole responsibility for a group of 20–50 undergraduates, three hours a week, for fifteen weeks—was for years nonexistent in my new life. When, rarely, I taught undergraduates, it was a group of 70 students, with a coteacher or a graduate assistant, four hours a week rather than three. Those differences may seem small, but they change the classroom dynamics significantly. For one thing, the arrangement here allows less flexibility during the semester; too many people are involved to change the syllabus midstream, whatever the class's interests and rhythms turn out to be. The inflexibility is not as total as in a medical school class, but it is substantial.

9. Andre, "Role Morality."

10. Ross, "Changing the HEC Mission."

CHAPTER THREE

1. The idea of this allegory arose from my hearing John Fletcher compare bioethics to an island between two continents: I believe he suggested that medicine or health care was one, and philosophy, or academia, the other.

2. This part of the story, incidentally, is little appreciated by clinicians in bioethics. Most academics who know what life has been like "on the market" stop trying to describe it to others. The brutality of medical training is well known, but the psychological violence of the "job wars" remains invisible. Years of competing for scarce jobs, not knowing whether you'll ever work again in your profession, knowing that next year's job will be hundreds or thousands of miles from this year's—this is like trying to describe being poor to those who never have been. They cannot understand, and if they respond at all it is in terms of pity or disdain. As a result, a certain unspoken bitterness persists. One philosopher bioethicist to whom I talked put it this way: "We're beginning to hear how hard it is for some young doctors, who go through subspecialty training and end up having to do something in primary care. I don't give a s——. This has been happening to our people forever."

3. Thomson, "A Defense of Abortion."

CHAPTER FOUR

1. X might follow from sheer logic, or from what is called conversational implicature: To say that X is real soy sauce is to imply that something else is not. This second sense is closer to the ordinary language sense of implication.

2. Clouser, "Philosophy, Literature, and Ethics"; Hawkins, "Literature, Philosophy, and Medical Ethics."

3. Todd, *Pidgins and Creoles,* 1–2.

4. Muhlhausler, *Pidgin and Creole Linguistics,* 5.

5. Mafeni, "Nigerian English," 95–96.

6. "Jargon" itself has a number of meanings, including a specialized one within linguistics. Here I mean it in the nontechnical, ordinary sense: "words or expressions developed for use within a particular group, hard for outsiders to understand." *Oxford American Dictionary,* 1980.

7. Glossary, *Journal of Pidgin and Creole Languages,* website. In November 2000 the Journal's website URL was http://www.siu.edu/departments/cola/ling/. Note that the entry refers to a jargon rather than a pidgin: in this context a jargon is a stage in the development of a pidgin when the vocabulary is small and its domain limited.

8. Cape York Creole for instance, has three forms of the first person plural. English offers only "we" (and "us"). The speaker of Cape York Creole can with a single word indicate whether the others for whom she speaks are present or absent, and whether there is one or more than one of them. In other words, there is a form of "we" that means "you [singular] and I," a form that means "you [plural] and I," another for "he or she and I," and one for "they and I." O'Grady and Dobrovolsky, *Contemporary Linguistics,* 348.

9. Clouser, *Teaching Bioethics,* 53–57.

10. Ibid., 55.

11. My thanks to Patricia Marshall for clarifying these points.

12. Capron, "What Contributions," 302.

13. Long, "Reflections on Becoming a Cucumber."

14. Drane, "Ethical Workup Guides Clinical Decision-Making."

15. Thomasma, "Training in Medical Ethics."

16. Wilmut et al., "Viable Offspring Derived."

CHAPTER FIVE

1. MacIntyre, *After Virtue,* 181–203.

2. Thomson, "A Defense of Abortion."

3. Beauchamp and Childress, *Principles,* vii.

4. Tom Stoppard, "Professional Foul," as quoted in Beauchamp and Childress, *Principles,* vii; Stoppard, *Every Good Boy.*

5. Beauchamp and Childress, *Principles,* vii.

6. As long ago as 1987, Jeremy Waldron commented of another moral philosopher, "Like almost everyone else who writes about moral philosophy these days, [he] makes the rather tiresome observation that almost everyone else who writes about moral philosophy these days thinks morality is a matter of the simple application of a single master rule." Waldron, "Review," 331.

7. Ruddick, "What Should We Teach," 21.

8. Ibid.

9. Walker, "Keeping Moral Space Open," 33.

10. Whitbeck, "Ethics as Design."

11. Fleck and Angell, "Please Don't Tell."

12. Letter of invitation to participants in the William Bennett Bean conference, February 1995. That invitation indirectly gave rise to this book.

13. Andre, "My Client, My Enemy."

14. Andre, Fleck, and Tomlinson, "Improving Our Aim."

15. Miller, "Virtues, Practices, and Justice," 250–52.

16. Kelly et al., "Understanding the Practice."

17. Henk ten Have made a comment to this effect during the working conference that culminated in Carson and Burns, *Philosophy of Medicine and Bioethics.*

CHAPTER SIX

1. MCW-Bioethics electronic list, August 6, 1997. MCW-Bioethics is run by the Medical College of Wisconsin. It is open only to bioethics professionals. To subscribe, send an e-mail message to listproc@post.its.mcw.edu stating, "Subscribe MCW-Bioethics [your first name and last name]." You will then be asked for further information about yourself.

2. Ibid., August 7, 1997.

3. Weisbard, "The Role of Philosophers"; Brock, "Truth or Consequences"; Kamm, "Philosopher as Insider and Outsider"; Benjamin, "Philosophical Integrity and Policy Development."

4. Society for Health and Human Values, *Core Competencies.*

5. Thomas, *Moral Development Theories.*

6. Ibid., 54; Piaget, *Moral Judgment of the Child.*

7. Kohlberg, *Philosophy of Moral Development.*

8. Gilligan, *In a Different Voice.*

9. Rest, "A Psychologist Looks at the Teaching of Ethics."

10. Among their more significant modifications, they no longer claim that the form taken by moral reasoning can be separated entirely from its content (e.g., they recognize that the belief that impartiality is a morally superior stance is itself a moral claim); they understand the moral world of the small child in much more complex ways; they think in terms of cognitive schemas and implicit knowledge, not just in terms of explicitly offered reasoning. Rest et al., *Postconventional Moral Thinking.*

11. Ibid., 100.

12. MacIntyre, *After Virtue;* Blum, *Friendship;* Williams, *Moral Luck.*

13. Pincoffs, *Quandaries and Virtues,* 30.

14. Ibid., 133.

15. Ibid., 174.

16. Blum, *Moral Perception.*

17. More specifically, from "concrete operational thinking" to "formal operational thinking" (e.g., to understanding the conservation of mass in spite of alterations in shape).

18. Blanchard-Fields, "Postformal Reasoning," 80.

19. Ibid.

20. Kramer, "Development of an Awareness of Contradiction," 141.

21. "Until the early modern period . . . 'believe' carried much the same range of meaning as that associated with 'to set the heart upon' . . . 'to hold dear' . . . 'to cherish.'" Fowler, *Stages of Faith,* 12. Fowler finds similar etymologies in German and Old English. In making these points, he is drawing from Smith, *Faith and Belief,* 76, 105–6.

22. I discuss the ways in which I believe philosophy should be learning more from religious scholarship in "Faith and the Unbelieving Ethics Teacher."

23. Fowler, *Stages of Faith,* 198.

24. Rest, "A Psychologist Looks at the Teaching of Ethics," 30.

25. Ibid., 29.

26. Ibid., 30.

27. For more on this point, see Rest, "Morality."

28. I explore some of this issue in "Learning to See."

29. Andre, "Humility Reconsidered."

30. On this point, see, for instance, Hill, "Servility and Self-Respect"; and Andre, "The Equal Moral Weight."

31. Oliner and Oliner, *Altruistic Personality.*

32. Rest, "Morality," 616.

33. Ibid., 568.

34. Among many critical treatments of medical school, see Konner, *Becoming a Doctor;* Klass, *A Not Entirely Benign Procedure;* Marion, *Learning to Play God;* and Shem, *House of God.*

35. Bosk, *Forgive and Remember.* See also Stewart, *Blind Eye.*

36. Suzanne Gordon, *Life Support.*

37. Pincoffs, *Quandaries and Virtues,* 16.

38. There has been a spate of work in the past ten years on the moral role of the emotions. One prominent figure has been Nussbaum in, among other writings, *Love's Knowledge.*

39. I first came across this point in Jenner and Gaylin, *Perversion of Autonomy.*

40. Aristotle's *Nicomachean Ethics* was originally part of a single book that included the *Politics.* I take the term "decent society" from Margalit, *Decent Society.*

41. Zagzebski, *Virtues of the Mind,* 137.

42. Ibid., 199, paraphrasing Braeton, "Towards a Feminist Reassessment."

43. Tomlinson and Czlonka, "Futility and Hospital Policy."

44. Andre, Fleck, and Tomlinson, "Improving Our Aim."

45. Health care as a benefit is untaxed compensation. As a result, those with the kind of job that offers the benefit pay less for health insurance (because it comes from untaxed dollars) than those who buy insurance out of pocket. If these benefits were taxed as income, the resulting staggering sum of money would pay for health insurance for every uninsured American.

46. Plenary session at the 1999 ASBH meeting; Reinhart, "Wanted," 1447.

CHAPTER SEVEN

1. Plenary session of a spring 1998 regional meeting of the Society for Health and Human Values, held at Youngstown State University.

2. London, "The Rescuers." Although this research is often cited, London is tentative about the conclusions he draws; his sample was small and his methodology casual.

3. Im, "Emotional Control and Virtue in the Mencius."

4. Zagzebski, *Virtues of the Mind,* 134–35.

5. There is a fair amount of discussion of this concept in the philosophical literature. I argue for a certain view of nature and grounding of special role obligations in "Role Morality."

6. Beatty, "Good Listening."

7. Jecker, Jonsen, and Pearlman, *Bioethics: An Introduction;* Jonsen, *Birth of Bioethics;* Rothman, *Strangers at the Bedside.*

8. Especially the first: National Commission, *The Belmont Report.*

9. For an excellent overview of the way law in the United States has shaped the field, see Capron, "What Contributions."

10. Strouse, "How to Give."

11. Chambers, "From the Ethicist's Point of View."

12. Gawande, "Under Suspicion."

13. See, for instance, Benjamin and Curtis, *Ethics in Nursing.*

14. Antonovsky, *Unraveling the Mystery of Health;* Syme, "Control and Health"; Syme, *Community Participation.*

15. Anda et al., "Depressed Affect," 285.

16. Hagan, "Defiance and Despair," 119–20.

17. Kawachi, Kennedy, and Wilkinson, *Income Inequality and Health.*

18. Ibid. The issue is now beginning to be addressed in bioethics; see Daniels, Kennedy, and Kawachi, "Why Justice Is Good."

19. Jameton, "Dilemmas of Moral Distress"; Wilkinson, "Moral Distress."

20. Battin, "Assisted Suicide."

21. Cassel, "Nature of Suffering," 639.

22. Cassel, "Recognizing Suffering," 25.

23. Jameton and Pierce, "Environment and Health."

24. Kuhn, *Structure of Scientific Revolutions.*

25. Executive Order 12975, October 3, 1995.

26. Alta Charo, plenary address, American Society for Bioethics and Humanities annual meeting, Houston, November 1998.

27. I don't know whether this is true or not. One person commented that the change is in bioethics fashion, and that the literature of the field has moved away from questions of unwanted overtreatment, but that it remains the most serious form of patient suffering.

28. Katz, *Silent World.*

29. The labels express various points of view about what counts as acceptable training. A direct-entry midwife is trained by other midwives and has no previous medical or nursing training.

30. Churchill, "The Ethicist in Professional Education"; Glazer, "Threat of the Stranger."

31. Simmel, "Sociological Significance," 322–27.

32. Ibid., 324.

33. Schuetz, "The Stranger," 500.

34. Tiryakian, "Sociological Perspectives on the Stranger," 53.

35. Collins, *Black Feminist Thought.*

36. Anzaldua and Moraga, *This Bridge Called My Back;* Lugones, "Purity, Impurity and Separation."

37. Liaschenko, "Artificial Personhood."

CHAPTER EIGHT

1. There is a substantial philosophic literature on integrity, but much less on discernment. See especially Martin Benjamin, *Splitting the Difference.* The rough summary I present here comes from personal conversation with him. The roots of the concept of discernment are in theology and religion.

2. For a useful overview, see Nichols, "Federal Science Policy and Universities."

3. Pearce, "Traditional Epidemiology."

4. Crossen, "Medical Researcher."

5. There was a covert suggestion that an "outside" doctor would use the information to help one of the girls, and this suggestion made a great impact on the family. Gordon and Bonkovsky, "Family Dynamics and Children."

6. Zoloth, "Audience and Authority," 356.

7. Ibid., 351.

8. Carl Elliott, "Miss Lonelyhearts."

CHAPTER NINE

1. Private communication.

2. Cherniss, *Beyond Burnout*, 140–50. He also found that certain personal characteristics help. One is an accurate sense of what one enjoys and can do well. Another is an ability to negotiate a complex system dispassionately, avoiding or resolving conflicts, and learning how to tap an institution's resources. It is also important to continue to learn, and to be serious about balancing work and home. In fact "the professionals who recovered . . . actually worked fewer hours" than those who did not (163). Finally, they had tempered their early expectations of themselves and set more realistic goals (151–68).

3. Ibid., 136–45, 180–90. Cherniss was studying human service professionals—counselors, high school teachers, and social workers. Most of our work is not as draining as what these professionals face, partly because for most of us autonomy, a flexible schedule, and the ability to learn (three of the characteristics he identified as crucial) are built into our jobs. His conclusion about the importance of a "shared moral vision" comes from his observation of a group of Catholic sisters, who avoided burnout in spite of having little autonomy or flexibility.

4. Colby and Damon, *Some Do Care*. They defined a moral exemplar as someone whose life would be admirable even to those who did not share his or her political and religious beliefs, who showed a sustained and deep commitment to certain ends throughout life, and who used tools and strategies that fit these fundamental commitments.

5. Ibid., 4.

6. Ibid., 4–5.

7. Ibid., 173.

8. Aristotle, *Nicomachean Ethics*, 1156b7–23.

9. Nichols, for instance, discusses the effects of funding within the sciences: "The campus, therefore, increasingly provides for many scientists only the base for a global intellectual life rather than a home for a meaningful institutional life." Nichols, "Federal Science Policy and Universities," 212.

10. The phrase became widely known through Robert Putnam's controversial work *Bowling Alone*.

11. Thirty years ago, as the second wave of twentieth-century feminism first grew strong, I hoped that work lives would evolve so that young parents typically shared both home and work, each holding perhaps a half- or two-thirds-time job, each contributing equally to child care and housework. Those part-time jobs I foresaw would have offered benefits and job security, proportionately. But nothing like this developed. See my discussion of the ever-busier work week later in this chapter for references to some economic and sociological analyses of the problem.

12. The contrast may be not only with philosophy but also with most of the humanities. A chaplain once remarked to me that her perspectives were far more welcomed in science departments than in the College of Arts and Letters.

13. This currently much-used word has many metaphorical meanings. I will stay with the simplest. For an interesting discussion of its many uses, current and otherwise, see Kleinman, *Writing at the Margin*.

14. In another contrast, articles reporting scientific findings almost always have many authors, so many that it can be a scandal. In some fields it has been typical to include everyone in the lab or even the institution, whether or not each person had anything to do with the project.

15. This is in sharp contrast to the heads of more traditional enterprises, like academic departments, among whom ability varies greatly. The difference probably results from the newness and uniqueness of bioethics, where only those with energy and vision will be able to start, or maintain, a center that is unlikely to fit the usual hospital or academic mold.

16. Gawande, "When Good Doctors Go Bad."

17. Schreuder, "Ascent into Hell." Krakauer's book about the experience is entitled *Into Thin Air*.

18. When I did similar things in California twenty years ago, we called it "careening around the county like a comet." That was inexact, though; the jobs were rarely all in the same county.

19. Zoloth, "Audience and Authority," 356.

20. The three associations that merged were the Society for Health and Human Values, the American Association of Bioethics, and the Society for Bioethics Consultation.

21. Society for Health and Human Values, *Core Competencies*.

22. "Publishing without Perishing: A Handbook for Graduate and Professional Students on Publishing in Bioethics and the Medical Humanities" [http://www.asbh.org/papers/].

23. "Prisoners' dilemmas" are situations in which everyone's pursuing his or her individual best interest leads to an outcome that is in no one's best interest. In this version, everyone's doing what seems individually justified perpetuates a situation that, as a whole, does not seem justified: that is, a situation in which professional travel funds are spent partially for play.

24. Kohn, "Managing People," 146. See also Kohn, *No Contest.*

25. In the mid-1990s, when a student scoring extraordinarily high on certain exams failed to win the medical school lottery, a few slots were set aside to be awarded on this kind of merit.

26. See the work of Schor, *Overworked American,* and the complementary analysis of Hochschild, *Time Bind.* Robinson and Godbey contest the claim that we are working longer hours, but agree that we feel more rushed in *Surprising Ways.*

CHAPTER TEN

1. Tolstoy, "The Death"; Starr, *Social Transformation;* Bosk, *Forgive and Remember.*

2. Beecher, "Ethics and Clinical Research."

3. National Commission, *Belmont Report;* President's Commission, *Deciding to Forego Life-Sustaining Treatment.* The President's Commission issued a number of reports, of which this has been one of the most influential.

4. Lynn and Childress, "Must Patients Always Be Given Food?"; Meisel, "Legal Myths."

5. One interesting candidate for that is Carl Elliott's *Slow Cures and Bad Philosophers,* which presents a significant new, Wittgensteinian approach to moral reflection.

6. One commentator, for instance, mapping the origins of the discipline of art history, speculates that in part it resulted from "the desire of university administrators to prepare future alumni to make capital investments in the institution—museums, campus architecture, archaeological excavations, and university travel bureaus." He also provides a useful overview of the varying significance of disciplinary affiliation internationally. Whereas in the United States, one's discipline has the strongest cultural power, in France, units center instead around method, shared intellectual values, and charismatic leaders. In Germany, "the major social unit . . . [is] the individual chair and its associated structures—the seminar and the research institute or laboratory." Reese, "Mapping Interdisciplinarity," 545, 546.

7. Mitchell, "Interdisciplinarity and Visual Culture," 540.

8. Ibid.

9. See, for instance, Linda Myrsiades, "Introduction: Law, Literature, and Interdisciplinarity," citing Stanley Fish.

10. Myrsiades, "Introduction: Law, Literature, and Interdisciplinarity," 1.

11. Dinmore, "Interdisciplinarity and Integrative Learning," 456 (emphasis added). Dinmore attributes the observation to Rustum Roy, "Interdisciplinary Science on Campus."

12. Parson, "Three Dilemmas," 316.

13. Ibid., 317.

14. MCW-Bioethics electronic list, 4 August 1997.

15. Myrsiades, "Interdisciplinarity, Law, Language, and Literature," 207–9.

16. Independent funding became a serious problem, and NABER has since dissolved.

17. Some other examples: In an earlier chapter I pointed out that conflicts between doctors and nurses, and the crises of conscience that result, go unremarked if one spends time only with doctors. But this is a problem for many other audiences: for medical students and residents, for therapists and social workers, and now for doctors as well, in their interface with managed care. Having responsibility without full authority is a general and powerful problem. Similarly, now that I spend time regularly in our veterinary school, I see questions of animal rights and animal welfare differently than when I only read the relevant philosophical literature. ("What do dogs really need?," asked my veterinarian colleague Sally Walshaw, talking about a new regulation requiring exercise for dogs used in research. "They don't need exercise. They need love." This points from still another angle to the general problem in research ethics, of developing through a political process rules that are actually useful.) Working with a graduate student from Africa on questions of confidentiality, I realized that American concepts and practices are not fully coherent on their own terms. For example, here, as well as in Zimbabwe, we often tell family before we tell the patient (especially frail patients, and patients just out of surgery), although our abstract theory would indicate otherwise. Our theory is probably mistaken. Or a different example: It is hard to defend the practice, here, as well as in Zimbabwe, of revealing information to everyone in a white coat, and no one without, so that an LPN on a different case knows about metastatic disease a week before the patient does. The problem is not so much ethnocentrism as a lack of self-reflection, but exposure to other ways of living and thinking helps with both.

18. When are we justified, for instance, in saying that it is raining, or the hat is black? It is not enough that we believe these things and that they be

true, since that might be a fortunate accident. So what kinds of justification—reasoning and evidence—change a true belief into knowledge? This is the problem as Plato presented it, and it is enormously more challenging than it seems. The systematic pursuit of the question eventually led to the realizations about the nature of knowledge that I summarize here.

19. Kvanvig, *Intellectual Virtues,* 183.

20. Ibid., 169.

21. Ibid.

22. Most of the recent work in philosophy addresses a question that began with Plato: When does a true belief become knowledge, rather than a lucky guess? It becomes knowledge when the knower has sufficient justification, which by the nature of things is usually short of absolute certainty. The work of the virtue epistemologists suggests that intellectual virtue should in some cases count as sufficient justification to make a true belief count as knowledge. For a useful summary, see Axtell, "Recent Work on Virtue Epistemology."

23. Zagzebski, *Virtues of the Mind,* 139.

24. Ibid., 157–58.

25. Ibid., 159.

26. Ibid.

27. Kvanvig, *Intellectual Virtues,* 194.

28. Ibid., 170–71.

29. Code, *Epistemic Responsibility,* 254.

30. Zagzebski, *Virtues of the Mind,* 158.

31. Code, *Epistemic Responsibility,* 57.

32. Ibid., 246.

33. The opening passage of the *Metaphysics:* "All men by nature desire to know. An indication of this is the delight we take in our senses; for even apart from their usefulness they are loved for themselves; and above all the sense of sight. For not only with a view to action, but even when we are not going to do anything, we prefer seeing (one might say) to everything else. The reason is that this, most of all the senses, makes us know" (980a23–27).

34. Kvanvig, *Intellectual Virtues,* 160.

35. Ibid., 192.

36. For other problems with the movement toward "accountability," see Strathern, *Audit Cultures.*

37. Her discussion is not only brave, it is wise, for she continues: "Still I am aware of the paradox of my power over these students. I am aware of my role, my place in an institution that is larger than myself, whose power I

wield even as I am powerless, whose shield of respectability shelters me even as I am disrespected." Patricia J. Williams, *Alchemy of Race and Rights,* 95–96.

38. The topic of the proper use of such evaluations is a separate and important one. That they can be helpful and informative is beyond doubt. That the custom sometimes allows administrators to discharge a responsibility without thought also seems obvious. "We take teaching seriously. We really look at student evaluations."

39. Leape, "Error in Medicine"; Hilfiker, *Healing the Wounds.*

40. This was Harold Rood, of Washburn University of Topeka, for whose kindness so many years ago I remain grateful.

41. Birnbaum, Newell, and Saxberg, "Managing Academic Interdisciplinary Research Projects," 645.

42. Code, *Epistemic Responsibility,* 56.

43. James, *The Will to Believe,* 30, as cited in Cooper, "Intellectual Virtues," 469.

44. Plenary session, ASBH, 1998.

45. The views he expressed that day appeared later in Cole-Turner, "Theological Perspectives on the Status of DNA."

46. See chapter 6, "Bioethics and Moral Development."

47. Cole, *Journey of Life.*

CHAPTER ELEVEN

1. Andre, Fleck, and Tomlinson, "Improving Our Aim."

2. Gallagher, "Religious Freedom."

3. Dresser, "Freedom of Conscience."

4. Andre, "Beyond Moral Reasoning."

5. Brody, *Placebos and the Philosophy of Medicine.*

6. Moore, "Autobiography," 14.

Bibliography

Anda, R., D. Williamson, D. Jones, C. Macera, E. Eaker, A. Glassman, and J. Marks. "Depressed Affect, Hopelessness, and the Risk of Ischemic Heart Disease in a Cohort of U.S. Adults." *Epidemiology* 4:4 (July 1993): 285–94.

Andre, Judith. "Beyond Moral Reasoning: A Wider View of the Professional Ethics Course." *Teaching Philosophy* 14:4 (December 1991): 359–73.

———. "The Equal Moral Weight of Self- and Other-Regarding Acts." *Canadian Journal of Philosophy* 17:1 (March 1987): 155–65.

———. "Faith and the Unbelieving Ethics Teacher." *Centennial Review* 34:2 (Spring 1990): 255–74.

———. "Goals of Ethics Consultation: Toward Clarity, Utility, and Fidelity." *Journal of Clinical Ethics* 8:2 (Summer 1997): 193–98.

———. "Humility Reconsidered." In *Margin of Error: The Ethics of Mistakes in the Practice of Medicine,* edited by Susan B. Rubin and Laurie Zoloth, 59–72. Hagerstown, Md.: University Publishing Group, 2000.

———. "Learning to See: Moral Growth during Medical School." *Journal of Medical Ethics* 18 (September 1992): 148–52.

———. "My Client, My Enemy." *Professional Ethics* 3:3–4 (1994): 27–46.

———. "Nagel, Williams, and Moral Luck." *Analysis* 43:4 (October 1983): 202–7.

———. "Role Morality as a Complex Instance of Ordinary Morality." *American Philosophical Quarterly* 28:1 (January 1991): 73–80.

Andre, Judith, Leonard Fleck, and Tom Tomlinson. "Improving Our Aim." *Journal of Medicine and Philosophy* 24:2 (April 1999): 130–47.

Antonovsky, Aaron. *Unraveling the Mystery of Health: How People Manage Stress and Stay Well.* San Francisco: Jossey-Bass, 1987.

Anzaldua, Gloria, and Cherrie Moraga, eds. *This Bridge Called My Back: Writings by Radical Women of Color.* Watertown, Mass.: Persephone Press, 1981.

Aristotle. *Metaphysics.* Translated by W. D. Ross. In *The Basic Works of Aristotle,* 689–926. New York: Random House, 1941.

———. *Nichomachean Ethics*. Translated by H. Rackman. Cambridge, Mass.: Harvard University Press, 1934.

Axtell, Guy. "Recent Work on Virtue Epistemology." *American Philosophical Quarterly* 34:1 (January 1997): 1–26.

Battin, Margaret P. "Assisted Suicide: Can We Learn from Germany?" *Hastings Center Report* 22:2 (March–April 1992): 44–51.

Beatty, Joseph. "Good Listening." *Educational Theory* 49:3 (Summer 1999): 281–98.

Beauchamp, Tom L., and James F. Childress. *Principles of Biomedical Ethics*. New York: Oxford University Press, 1979.

Beecher, Henry K. "Ethics and Clinical Research." *New England Journal of Medicine* 74 (June 16, 1966): 1354–60.

Benjamin, Martin. "Philosophical Integrity and Policy Development in Bioethics." *Journal of Medicine and Philosophy* 15:4 (August 1990): 375–89.

———. *Splitting the Difference: Compromise and Integrity in Ethics and Politics*. Lawrence: University of Kansas Press, 1990.

Benjamin, Martin, and Joy Curtis. *Ethics in Nursing*. New York: Oxford University Press, 1992.

Birnbaum, Philip H., William T. Newell, and Borje O. Saxberg. "Managing Academic Interdisciplinary Research Projects." *Decision Sciences* 10:4 (October 1979): 645–63.

Blanchard-Fields, Fredda. "Postformal Reasoning in a Socioemotional Context." In *Adult Development*, vol. 1: *Comparisons and Applications of Developmental Models*, edited by Michael L. Commons, Jan D. Sinnott, Francis A. Richards, and Cheryl Armon, 73–94. New York: Praeger, 1989.

Blum, Lawrence A. *Friendship, Altruism, and Morality*. Boston: Routledge & Kegan Paul, 1980.

———. *Moral Perception and Particularity*. New York: Cambridge University Press, 1994.

Bosk, Charles L. *Forgive and Remember: Managing Medical Failure*. Chicago: University of Chicago Press, 1979.

Braeton, Jane. "Towards a Feminist Reassessment of Intellectual Virtue." *Hypatia* 5:3 (Fall 1990): 1–14.

Brock, Dan. "Truth or Consequences: The Role of Philosophers in Policy-Making." *Ethics* 97:4 (July 1987): 786–91.

Brody, Howard. *Placebos and the Philosophy of Medicine: Clinical, Conceptual, and Ethical Issues*. Chicago: University of Chicago Press, 1980.

Capron, Alexander Morgan. "What Contributions Have Social Science and the Law Made to the Development of Policy on Bioethics?" *Daedalus* 128:4 (Fall 1999): 295–325.

Carson, Ronald A., and Chester R. Burns, eds. *Philosophy of Medicine and Bioethics: A Twenty-Year Retrospective and Critical Appraisal.* Boston: Kluwer Academic Publishers, 1997.

Cassel, Eric J. "The Nature of Suffering and the Goals of Medicine." *New England Journal of Medicine* 36:11 (March 1982): 639–45.

———. "Recognizing Suffering." *Hastings Center Report* 21:3 (May–June 1991): 24–31.

Chambers, Tod. "From the Ethicist's Point of View." *Hastings Center Report* 26:1 (January–February 1996): 25–32.

Cherniss, Cary. *Beyond Burnout: Helping Teachers, Nurses, Therapists, and Lawyers Recover from Stress and Disillusionment.* New York: Routledge, 1995.

Churchill, Larry. "The Ethicist in Professional Education." *Hastings Center Report* 8:6 (December 1978): 13–15.

Clouser, K. Danner. "Philosophy, Literature, and Ethics: Let the Engagement Begin." *Journal of Medicine and Philosophy* 21 (June 1996): 321–40.

———. *Teaching Bioethics: Strategies, Problems, and Resources.* New York: Hastings Center, 1980.

Code, Lorraine. *Epistemic Responsibility.* Hanover, N.H.: University Press of New England, 1987.

Colby, Anne, and William Damon. *Some Do Care: Contemporary Lives of Moral Commitment.* New York: Free Press, 1992.

Cole, Thomas R. *The Journey of Life: A Cultural History of Aging in America.* New York: Cambridge University Press, 1992.

Cole-Turner, Ronald. "Theological Perspectives on the Status of DNA: A Contribution to the Debate over Genetic Patenting." In *Perspectives on Genetic Patenting,* edited by Audrey R. Chapman, 149–65. Washington, D.C.: American Association for the Advancement of Science, 1999.

Collins, Patricia Hill. *Black Feminist Thought: Knowledge, Consciousness, and the Politics of Empowerment.* New York: Routledge, 1991.

Cooper, Neil. "The Intellectual Virtues." *Philosophy* 69 (1994): 459–69.

Crossen, Cynthia. "A Medical Researcher Pays for Challenging Drug-Industry Funding." *Wall Street Journal* (January 3, 2001), sect. A, p. 1.

Daniels, Norman, Bruce P. Kennedy, and Ichiro Kawachi. "Why Justice Is Good for Our Health: The Social Determinants of Health Inequalities." *Daedalus* 128:4 (Fall 1999): 215–51.

Dinmore, Ian. "Interdisciplinarity and Integrative Learning: An Imperative for Adult Education." *Education* 117 (Spring 1997): 452–67.

Drane, James F. "Ethical Workup Guides Clinical Decision-Making." *Health Progress* 69:11 (December 1988): 64–67.

Dresser, Rebecca S. "Freedom of Conscience, Professional Responsibility, and Access to Abortion." *Journal of Law, Medicine and Ethics* 22:3 (Fall 1994): 280–85.

Elliott, Carl. "Miss Lonelyhearts." In *Practicing the Medical Humanities: Forms of Engagement*, edited by Ronald Carson, Chester R. Burns, and Thomas R. Cole. Hagerstown, Md.: University Publishing Group, forthcoming.

———. *Slow Cures and Bad Philosophers: Essays on Wittgenstein, Medicine, and Bioethics.* Durham, N.C.: Duke University Press, forthcoming.

Executive Order 12975. *Presidential Documents.* Federal Register, 60:193 (October 5, 1995): 520063–65. [http://bioethics.gov/about/eo12975.html]

Fleck, Len, and Marcia Angell. "Please Don't Tell: Case Study and Commentaries." *Hastings Center Report* 21:6 (November–December 1991): 39–40.

Fowler, James W. *Stages of Faith: The Psychology of Human Development and the Quest for Meaning.* San Francisco: Harper and Row, 1981.

Gallagher, Janet. "Religious Freedom, Reproductive Health Care, and Hospital Mergers." *JAMWA* 52:2 (Spring 1997): 65–68, 84.

Gawande, Atul. "Under Suspicion." *New Yorker* 76:42 (January 8, 2001): 50–53.

———. "When Good Doctors Go Bad." *New Yorker* 76:22 (August 7, 2000): 60–66.

Gilligan, Carol. *In a Different Voice: Psychological Theory and Women's Development.* Cambridge, Mass.: Harvard University Press, 1982.

Gladwell, Malcolm. "Do Parents Matter." *New Yorker* 74:24 (August 17, 1998): 54–64.

Glazer, Myron. "The Threat of the Stranger." *Hastings Center Report* (October 1980): 25–31.

Gordon, Suzanne. *Life Support: Three Nurses on the Front Line.* Boston: Little, Brown, 1997.

Gordon, Valery M., and Frederick O. Bonkovsky. "Family Dynamics and Children in Medical Research." *Journal of Clinical Ethics* 7 (Winter 1996): 349–54.

Hagan, John. "Defiance and Despair: Subcultural and Structural Linkages between Delinquency and Despair in the Life Course." *Social Forces* 76:1 (September 1997): 119–235.

Hart, Nicky. *The Sociology of Health and Medicine.* Ormskirk, U.K.: Causeway Press, 1985.

Hawkins, Anne Hunsaker. "Literature, Philosophy, and Medical Ethics: Let the Dialogue Go On." *Journal of Medicine and Philosophy* 21 (June 1996): 341–54.

Henderson, David K. "Epistemic Competence and Contextualist Epistemology: Why Contextualism Is Not Just the Poor Person's Coherentism." *Journal of Philosophy* 91:12 (1994): 627–49.

Hilfiker, David. *Healing the Wounds.* New York: Penguin, 1987.

———. *Not All of Us Are Saints.* New York: Hill and Wang, 1994.

Hill, Thomas. "Servility and Self-Respect." *Monist* 57:1 (January 1973): 87–104.

Hochschild, Arlie Russell. *The Time Bind: When Work Becomes Home and Home Becomes Work.* New York: Metropolitan Books, 1997.

Im, Manyul. "Emotional Control and Virtue in the Mencius." *Philosophy East and West* 49:1 (January 1999): 1–27.

James, William. *The Will to Believe and Other Essays in Popular Philosophy.* London: Longmans, Green, 1912.

Jameton, A. "Dilemmas of Moral Distress: Moral Responsibility and Nursing Practice." *American Organization of Women's Health and Neonatal Nurses* 4:4 (1993): 542–51.

Jameton, A., and Jessica Pierce. "Environment and Health: 8. Sustainable Health Care and Emerging Ethical Responsibilities." *Canadian Medical Association Journal* 164:3 (2001): 365–69.

Jecker, Nancy S., Albert R. Jonsen, and Robert A. Pearlman. *Bioethics: An Introduction to the History, Methods, and Practice.* Sudbury, Mass.: Jones and Barlett, 1997.

Jenner, Bruce, and Willard Gaylin. *The Perversion of Autonomy: The Proper Uses of Coercion and Constraints in a Liberal Society.* New York: Free Press, 1996.

Jonsen, Albert R. *The Birth of Bioethics.* New York: Oxford University Press, 1998.

Kamm, Frances. "The Philosopher as Insider and Outsider." *Business and Professional Ethics Journal* 9 (1990): 7–20.

Katz, Jay. *The Silent World of Doctor and Patient.* New York: Free Press, 1984.

Kawachi, Ichiro, Bruce P. Kennedy, and Richard G. Wilkinson. *Income Inequality and Health.* Vol. 1 of *The Society and Population Health Reader.* New York: New Press, 1999.

Kelly, Susan E., Patricia A. Marshall, Lee M. Sanders, Thomas A. Raffin, and Barbara A. Koenig. "Understanding the Practice of Ethics Consultation:

Results of an Ethnographic Multi-site Study." *Journal of Clinical Ethics* 8:2 (Summer 1997): 136–49.

Klass, Perri. *A Not Entirely Benign Procedure: Four Years as a Medical Student.* New York: Putnam, 1987.

Kleinman, Arthur. *Writing at the Margin.* Berkeley: University of California Press, 1995.

Kohlberg, Lawrence. *The Philosophy of Moral Development: Moral Stages and the Idea of Justice.* San Francisco: Harper and Row, 1981.

Kohn, Alfie. "Managing People: No Contest." *Inc.* 9:12 (November 1987): 145–48.

———. *No Contest: The Case against Competition.* Boston: Houghton Mifflin, 1992.

Konner, Melvin. *Becoming a Doctor: A Journey of Initiation in Medical School.* New York: Viking, 1987.

Krakauer, Jon. *Into Thin Air: A Personal Account of the Mount Everest Disaster.* New York: Villard, 1997.

Kramer, Deirdre A. "Development of an Awareness of Contradiction across the Life Span and the Question of Postformal Operations." In *Adult Development,* vol. 1: *Comparisons and Applications of Developmental Models,* edited by Michael L. Commons, Jan D. Sinnott, Francis A. Richards, and Cheryl Armon, 133–60. New York: Praeger, 1989.

Kuhn, Thomas S. *The Structure of Scientific Revolutions.* Chicago: University of Chicago Press, 1962.

Kvanvig, Jonathan L. *The Intellectual Virtues and the Life of the Mind.* Savage, Md.: Rowman and Littlefield, 1992.

Leape, Lucien L. "Error in Medicine." *Journal of the American Medical Association* 272:23 (December 1994): 1851–57.

Liaschenko, Joan. "Artificial Personhood: Nursing Ethics in a Medical World." *Nursing Ethics* 2:3 (September 1995): 185–96.

London, Perry. "The Rescuers: Motivational Hypotheses about Christians Who Saved Jews from the Nazis." In *Altruism and Helping Behavior: Social Psychological Studies of Some Antecedents and Consequences,* edited by J. Macaulay and L. Berkowitz, 241–50. New York: Academic Press, 1970.

Long, Susan Orpett. "Reflections on Becoming a Cucumber: Images of the Good Death in Japan and the United States." Unpublished manuscript.

Lugones, Maria. "Purity, Impurity and Separation." *Signs* 19:21 (Winter 1994): 458–79.

Lukes, Steven. *Power.* London: Macmillan, 1974.

Lynn, Joanne, and James F. Childress. "Must Patients Always Be Given Food and Water?" *Hastings Center Report* 13:5 (October 1983): 17–21.

MacIntyre, Alasdair. *After Virtue: A Study in Moral Theory.* Notre Dame, Ind.: University of Notre Dame Press, 1984.

Mafeni, Bernard. "Nigerian English." In *The English Language in West Africa,* edited by J. Spencer, 95–112. London: Longman, 1971.

Margalit, Avishai. *The Decent Society.* Cambridge, Mass.: Harvard University Press, 1996.

Marion, Robert. *Learning to Play God: The Coming of Age of a Young Doctor.* New York: Fawcett Crest, 1993.

Meisel, Alan. "Legal Myths about Terminating Life Support." *Archives of Internal Medicine* 151:8 (August 1991): 1497–1502.

Miller, David. "Virtues, Practices, and Justice." In *After MacIntyre,* edited by John Horton and Susan Mendus, 245–64. Notre Dame, Ind.: University of Notre Dame Press, 1994.

Mitchell, W. J. T. "Interdisciplinarity and Visual Culture." *Art Bulletin* 77 (December 5, 1995): 540–44.

Moore, G. E. "An Autobiography." In *The Philosophy of G. E. Moore,* edited by Paul A. Schilpp, 1–39. New York: Tudor Publishing Company, 1952.

Muhlhausler, Peter. *Pidgin and Creole Linguistics.* Oxford: Basil Blackwell, 1986.

Myrsiades, Linda. "Interdisciplinarity, Law, Language, and Literature." *College Literature* 23:1 (February 1996): 204–18.

———. "Introduction: Law, Literature, and Interdisciplinarity." *College Literature* 25:1 (Winter 1998): 1–11.

National Commission for the Protection of Human Subjects of Biomedical and Behavioral Research. *The Belmont Report: Ethical Principles and Guidelines for the Protection of Human Subjects of Research.* Washington, D.C.: Government Printing Office, 1978.

Nichols, Rodney W. "Federal Science Policy and Universities: Consequences of Success." *Daedalus* 122:4 (Fall 1993): 197–224.

Nussbaum, Martha. *Love's Knowledge: Essays on Philosophy and Literature.* New York: Oxford University Press, 1990.

O'Grady, William, and Michael Dobrovolsky. *Contemporary Linguistics: An Introduction.* New York: St. Martin's, 1989.

Oliner, Samuel P., and Pearl M. Oliner. *The Altruistic Personality.* New York: Free Press, 1988.

Parson, Edward A. "Three Dilemmas in the Integrated Assessment of Climate Change." *Climatic Change* 34:3–4 (November 1996): 315–26.

Pearce, Neil. "Traditional Epidemiology, Modern Epidemiology, and Public Health." *American Journal of Public Health* 86:5 (May 1996): 678–83.

Piaget, Jean. *The Moral Judgment of the Child.* New York: Free Press, 1965.

Pincoffs, Edmund L. *Quandaries and Virtues: Against Reductivism in Ethics.* Lawrence: University Press of Kansas, 1986.

President's Commission for the Study of Ethical Problems in Medicine and Biomedical and Behavioral Research. *Deciding to Forego Life-Sustaining Treatment: A Report on the Ethical, Medical, and Legal Issues in Treatment Decisions.* Washington, D.C.: Government Printing Office, 1983.

Putnam, Robert D. *Bowling Alone: The Collapse and Revival of American Community.* New York: Simon and Schuster, 2000.

Reese, Thomas F. "Mapping Interdisciplinarity." *Art Bulletin* 77:4 (December 1995): 544–49.

Reinhart, Uwe E. "Wanted: A Clearly Articulated Social Ethic for American Health Care." *JAMA* 278:17 (November 1997): 1446–47.

Reiser, Stanley Joel. "The Ethical Life of Health Care Organizations." *Hastings Center Report* 24:6 (November–December 1994): 28–35.

Rest, James. "Morality." In *Handbook of Child Psychology,* edited by Paul H. Mussen, vol. 3: *Cognitive Development,* edited by John H. Flavell and Ellen M. Markman, 556–629. New York: John Wiley and Sons, 1983.

———. "A Psychologist Looks at the Teaching of Ethics: Moral Development and Moral Education." *Hastings Center Report* 12:1 (February 1982): 29–36.

Rest, James, Darcia Narvaez, Muriel J. Bebeau, and Stephen J. Thomas. *Postconventional Moral Thinking: A Neo-Kohlbergian Approach.* Mahwah, N.J.: Lawrence Erlbaum Association, 1999.

Richardson, Henry S. "Specifying Norms as a Way to Resolve Concrete Ethical Problems." *Philosophy and Public Affairs* 19:4 (Autumn 1990): 279–310.

Robinson, John P., and Geoffrey Godbey. *The Surprising Ways Americans Use Their Time.* University Park: Pennsylvania State University Press, 1997.

Ross, Judith Wilson. "Changing the HEC Mission." *HEC (Hospital Ethics Committee) Forum* 12:1 (March 2000): 4–7.

Rothman, David J. *Strangers at the Bedside: A History of How Law and Bioethics Transformed Medical Decision Making.* New York: Basic Books, 1991.

Roy, Rustum. "Interdisciplinary Science on Campus: The Elusive Dream." In *Interdisciplinarity and Higher Education,* edited by J. Kockelmans, 161–96. University Park: Pennsylvania State University Press, 1979.

Ruddick, William. "What Should We Teach and Test?" *Hastings Center Report* 13:3 (June 1983): 20–22.

Schor, Juliet. *The Overworked American: The Unexpected Decline of Leisure.* New York: Basic Books, 1991.

Schreuder, Cindy. "Ascent into Hell: Mt. Everest Survivor Jon Krakauer Asks the Big Question: Why Me?" *Chicago Tribune,* North Sports Final, CN Edition (May 30, 1997): 1.

Schuetz, Alfred. "The Stranger: An Essay in Social Psychology." *American Journal of Sociology* 49:6 (May 1944): 499–507.

Shem, Samuel. *The House of God.* New York: Putnam, 1978.

Simmel, Georg. "The Sociological Significance of the 'Stranger.'" In *Introduction to the Science of Sociology,* edited by Robert E. Park and Ernest W. Burgess, 322–26. Chicago: University of Chicago Press, 1921.

Smith, Wilfred Cantwell. *Faith and Belief.* Princeton, N.J.: Princeton University Press, 1979.

Society for Health and Human Values–Society for Bioethics Consultations Task Force on Standards for Bioethics Consultation. *Core Competencies for Health Care Ethics Consultation.* Glenview, Ill.: American Society for Bioethics and Humanities, 1998.

Starr, Paul. *Social Transformation of American Medicine.* New York: Basic Books, 1982.

Stewart, James B. *Blind Eye: How the Medical Establishment Let a Doctor Get Away with Murder.* New York: Simon and Schuster, 1999.

Stoppard, Tom. *Every Good Boy Deserves Favor and Professional Foul: Two Plays.* New York: Grove Press, 1988.

Strathern, Marilyn. *Audit Cultures: Anthropological Studies in Accountability, Ethics and the Academy.* New York: Routledge, 2000.

Strouse, Jean. "How to Give Away $21.8 Billion." *New York Times Magazine* (April 16, 2000): 56–63, 78, 88, 96, 101.

Syme, S. Leonard. *Community Participation, Empowerment, and Health: Development of a Wellness Guide for California.* California Wellness Lecture Series, October 29, 1997. N.p.: California Wellness Foundation, University of California, 1997.

———. "Control and Health: An Epidemiological Perspective." In *Self Directedness: Cause and Effects throughout the Life Course,* edited by Judith Rodin, Carmi Schooler, and K. Warner Shaie, 213–29. Hillsdale, N.J.: Lawrence Erlbaum Association, 1990.

Thomas, R. Murray. *Moral Development Theories—Secular and Religious: A Comparative Study.* Westport, Conn.: Greenwood Press, 1997.

Thomasma, David. "Training in Medical Ethics: An Ethical Workup." *Forum on Medicine* 1:12 (1978): 33–36.

Thomson, Judith Jarvis. "A Defense of Abortion." In *Ethics and Public Policy: An Introduction to Ethics,* edited by Tom L. Beauchamp and Terry P. Pinkard, 268–83. Englewood Cliffs, N.J.: Prentice-Hall, 1983.

Tiryakian, Edward A. "Sociological Perspectives on the Stranger." *Soundings* 65:1 (1973): 45–58.

Todd, Loreto. *Pidgins and Creoles.* London: Routledge Press, 1974.

Tolstoy, Leo. "The Death of Ivan Ilych." In *The Death of Ivan Ilych, and Other Stories.* New York: New American Library, 1960.

Tomlinson, Tom, and Diane Czlonka. "Futility and Hospital Policy." *Hastings Center Report* 25:3 (May–June 1995): 28–35.

Toulmin, Stephen. "How Medicine Saved the Life of Ethics." *Perspectives in Biology and Medicine* 25:4 (Summer 1982): 736–50.

Waldron, Jeremy. Review of *Patterns of Moral Complexity,* by Charles Larmore. *Journal of Philosophy* 86:6 (June 1989): 331–33.

Walker, Margaret Urban. "Keeping Moral Space Open." *Hastings Center Report* 23:2 (March–April 1993): 33–40.

Webster, Chester, ed. *Caring for Health: History and Diversity.* Buckingham, Eng.: Open University, 1985.

Weisbard, Alan J. "The Role of Philosophers in the Public Policy Process: A View from the President's Commission." *Ethics* 97:4 (July 1987): 776–85.

Whitbeck, Caroline. "Ethics as Design: Doing Justice to Moral Problems." *Hastings Center Report* 26:3 (May–June 1996): 9–16.

Wilkinson, J. M. "Moral Distress: A Labor and Delivery Nurse's Experience." *Journal of Obstetrical and Gynecological Nursing* (November–December 1994): 513–21.

Williams, Bernard. *Moral Luck.* New York: Cambridge University Press, 1981.

Williams, Patricia J. *The Alchemy of Race and Rights.* Cambridge, Mass.: Harvard University Press, 1991.

Wilmut, I., A. E. Schnieke, J. McWhir, A. J. Kind, and K. H. S. Campbell. "Viable Offspring Derived from Fetal and Adult Mammalian Cells." *Nature* 385 (February 27, 1997): 810–13.

Zagzebski, Linda Trinkaus. *Virtues of the Mind: An Inquiry into the Nature of Virtue and the Ethical Foundations of Knowledge.* Cambridge: Cambridge University Press, 1996.

Zoloth, Laurie. "Audience and Authority: The Story in Front of the Story." *Journal of Clinical Ethics* 7 (Winter 1996): 355–61.

Index

STUDIES IN SOCIAL MEDICINE

Nancy M. P. King, Gail E. Henderson, and Jane Stein, eds.,
Beyond Regulations: Ethics in Human Subjects Research (1999).

Laurie Zoloth,
Health Care and the Ethics of Encounter:
A Jewish Discussion of Social Justice (1999).

Susan M. Reverby, ed.,
Tuskegee's Truths: Rethinking the Tuskegee Syphilis Study (2000).

Beatrix Hoffman,
Wages of Sickness:
The Politics of Health Insurance in Progressive America (2000).

Margarete Sandelowski,
Devices and Desires:
Gender, Technology, and American Nursing (2000).

Keith Wailoo,
Dying in the City of the Blues: Sickle Cell Anemia and the
Politics of Race and Health (2001).

Judith Andre,
Bioethics as Practice (2002).